Prinz-Albert-Studien

Prince Albert Studies

Prinz-Albert-Studien

Prince Albert Studies

Herausgegeben von / edited by

Franz Bosbach

Band 24 / Volume 24

Umwelt und Geschichte in Deutschland und Großbritannien

Environment and History in Britain and Germany

Herausgegeben von
Franz Bosbach, Jens Ivo Engels und Fiona Watson

K · G · Saur München 2006

Gedruckt mit Unterstützung der Universität Bayreuth

Bibliografische Information Der Deutschen Nationalbibliothek
Die Deutsche Nationalbibliothek verzeichnet diese Publikation
in der Deutschen Nationalbibliografie; detaillierte bibliografische Daten
sind im Internet über ***http://dnb.d-nb.de*** abrufbar.

∞

Gedruckt auf säurefreiem Papier / Printed on acid-free paper

Satz / Typesetting by Textservice Zink, Schwarzach
Druck und Bindung / Printed and bound by Strauss GmbH, Mörlenbach
Printed in Germany

ISSN 0941-6242 (Prinz-Albert-Studien)
ISBN-13: 978-3-598-21424-0 · ISBN-10: 3-598-21424-3

Inhalt/Contents

Abbildungsverzeichnis/List of Illustrations

Vorwort

Am 9. und 10. September 2005 fand im Riesensaal von Schloss Ehrenburg in Coburg die 24. Konferenz der Prinz-Albert-Gesellschaft statt. Unter dem Thema „Umwelt und Geschichte in Deutschland und Großbritannien – Environment and History in Britain and Germany" trugen deutsche und britische Historiker Ergebnisse zu dem noch jungen Forschungsfeld zum ersten Mal im Ländervergleich vor. Bei der Planung und Durchführung der Tagung sowie bei der Drucklegung kooperierten Fiona Watson von dem AHRC Research Centre for Environmental History in Stirling und Jens Ivo Engels vom Lehrstuhl für Wirtschafts- und Sozialgeschichte der Universität Freiburg i.Br., denen dafür ein besonderer Dank gilt.

In diesem Band werden die Texte der Konferenzbeiträge abgedruckt.

Die Tagung konnte durchgeführt werden dank der großzügigen Unterstützung, die die Prinz-Albert-Gesellschaft durch die Universität Bayreuth, den Universitätsverein Bayreuth, die Stadt Coburg, die Niederfüllbacher Stiftung sowie die Sparkasse Coburg und Lichtenfels erfahren hat. Ihnen allen sei dafür herzlich gedankt.

Ebenso ist denjenigen zu danken, die bei der Organisation der Tagung und bei der Drucklegung dieses Bandes behilflich waren: Katharina Beiergrößlein M.A., Dr. Thomas Brockmann, Claudia Ficht, Dagmar Hanke, Dirk Pfeifer sowie Andreas Brandmair vom K.G. Saur-Verlag in München.

Bayreuth, im Mai 2006

Franz Bosbach

Fiona Watson – Jens Ivo Engels

Introduction

It seems fitting, given Prince Albert's concern with agriculture and the influence that he and his wife had in popularising the highlands of Scotland, one of the wildest and most scenic areas of the United Kingdom, that the Prince Albert Society should host a conference to examine the state of Environmental History in Great Britain and Germany. This is also a most welcome opportunity to present this comparatively new area of study, in European terms, to a mainstream historical audience.

Reflections on the underlying principles and goals of Environmental History

The guiding principle underpinning Environmental History is to embed nature within the more usual concerns of the discipline. It aims to understand both how the environment, at all levels, has shaped and influenced the human condition and how humans have themselves related to and altered the environment. Environmental historians no longer aspire to telling the story of human history in an entirely new way, as was argued two decades ago. Nor is Environmental History natural history – the story of non-human nature. Rather, it is the study of the interrelations between humans and non-human nature in the past, placing the landscape and the environment back into the story of our own species where, in reality, it has always been.

However, as with other aspects of history, Environmental History also reflects the concerns of its contemporary practitioners, not least because, in many parts of the world, it has developed out of colonial or imperial history and what might be termed a 'clash of species' as well as civilisations. Classically, the subject has encompassed three main dimensions[1]. First, attempts to understand the major trends in particular aspects of change in the environment over time, such as climate and landuse history. Second, and relatedly, the strategies devised by humans to deal with such change. This might be termed the institutional aspect of Environmental History, since it usually concerns legislative or bureaucratic reactions to problems, whether by the state or non-government organisations. The cultural representation of nature has also oc-

[1] *J.R. McNeill*, Observations on the Nature and Culture of Environmental History, in: History and Theory, Theme Issue 42 (2003), 5-43.

cupied the attention of many historians, ranging from broadbrush, landscape-scale analyses to much more detailed, site-specific and/or species-oriented studies. More recently, interest has developed in the history of co-evolution (animal history) or of limited or serial confrontation between humans and nature (disaster history).

Given the extent to which the environment is interwoven into human history (and vice-versa), it is not surprising that Environmental History as a subject is so diverse and appears, at times, to be infeasibly all-encompassing. While ecology has been highly influential in the development of the subject, its limitations in terms of describing the complexity of human society have already been reached. Its successor – the concept of energy regimes, which stems from the idea of ecological footprints – has also been criticised for its general impracticality and – again – over-simplification of human systems. The truth of the matter remains that environmental historians have not yet developed a central methodological framework which would allow them to describe the relationship between humans and nature coherently. Nevertheless, it remains to be seen whether a degree of methodological coherence can be achieved whilst maintaining the complex analyses of human action which is traditionally the preserve of historians.

Finally, environmental historians, like philosophers, must also wrestle with another crucial issue in human-nature relations, as articulated by Professor Joachim Radkau[2]. Nature is, on the one hand, profoundly shaped by human activity and, in this respect, can be regarded as an artefact to be studied from a range of perspectives, such as scientific, technical, cultural, and philosophical. On the other hand, nature also exists beyond human experience and in its own right. However, as with the quantum physics problem of Schroedinger's cat, it is impossible to 'measure' or 'describe' this untouched nature, because the minute that you do, you have altered it, in this case by bringing to bear the individual subjective perspective.

State of the art in Environmental History in Great Britain and Germany

As already mentioned, Environmental History as an academic subject encompasses an extremely wide range of topics and approaches. Recently there has been an attempt to develop the beginnings of a structural framework, including a debate on methodology and also chronology, through the issue of 'turning points'[3]. However, any such attempt highlights another important point about Environmental History, namely the historiographical differences emerging from the various geocultural perspectives of its practitioners. In the United States which, along with Australia, began the formal study of the subject over thirty years ago, the issue of wilderness has been

2 *Joachim Radkau*, Was ist Umweltgeschichte?, in: Umweltgeschichte, *Werner Abelshauser* ed., Göttingen 1994, 11-28.

3 The Turning Points of Environmental History, *Joachim Radkau* and *Frank Uekötter* eds., Lanham, forthcoming.

a major topic of debate, simply because the country's vast, apparently untouched, landscapes have made such an impact on the national culture. There is also an explicit connection between American Environmental History and environmentalism, certainly among some of its early and leading practitioners such as Donald Worster, a relationship that would not necessarily be endorsed elsewhere.

A decade or so later, historians of the 'Old World' began to take notice of the environment, predominantly through an awareness of its growing role in Imperial and Third World History. This led to an appreciation of the impact which the experience of living and working in vastly different climates and observing exotic ecosystems had on intellectual thought and scientific endeavour back home[4]. In this context, we are therefore able to trace the influence of colonial Britain, particularly in forestry and dessication theories. German scientific approaches were borrowed and adapted for the Indian colonial experience and ultimately brought back to Britain with some rather bizarre consequences. Thus techniques designed to prevent dessication of the soil managed to influence forestry in Scotland, whose climate is renowned for the presence, rather than absence, of rainfall.

Finally, Environmental History of the 'Old World' itself also emerged as a distinct subject of study. This development has been most effectively acknowledged and supported by the founding of the European Society for Environmental History in 1999. The Society's activities and influence has allowed scholars of European environmental history to view their work within a much more international context[5].

However, despite the existence of European environmental historians across the continent and beyond, it would be true to say that the subject is probably at its most vibrant and coherent in Great Britain and Germany. Nevertheless, as Professors Smout and Brüggemeier's papers clearly indicate, the historiographical roots – and therefore primary concerns – of Environmental History in each country are as different from each other as they are from those of the 'New World' or elsewhere.

In essence, German Environmental History has evolved out of traditional elements of history, rather than from the sciences or social sciences, as is the case in Great Britain. As such, its sources and methods also derive almost exclusively from those classically used in History. Nevertheless, in its first two decades, Environmental History in Germany developed as a very specialized and discrete sub-discipline. However, environmental historians have recently worked hard to connect their research with other branches of history. As a result, they have explored the interrelations between Environmental History and the history of infrastructures, the history of disasters, agrarian history, the history of science, political and cultural history, the

4 See, for example, *Alfred W. Crosby*, Ecological Imperialism: the Biological Expansion of Europe, 900-1900, Cambridge 1986; Conservation in Africa, People, Policies and Practice, *D. Anderson* and *Richard Grove* eds., Cambridge 1987; *John M. MacKenzie*, Imperialism and the Natural World, Manchester 1990; *Richard H. Grove*, Green Imperialism: Colonial Expansion, Tropical Island Edens and the Origins of Environmentalism, 1600-1800, Cambridge and New Delhi 1995.

5 www.eseh.org.

history of power etc.[6] This has allowed Environmental History to enter the 'mainstream' of the discipline, to some degree at least. The papers featured in this book illustrate this point particularly well, showing some of the main points of contact with other fields of historical research. In essence, resource use is a matter of power relations; urban development in the 19th and 20th centuries was strongly influenced by the management of environmental factors; and, the protection of the environment reveals important insights into the function of (different) political systems.

German historians were originally persuaded to take notice of the environment, over twenty years ago, as an extension of more mainstream historical concerns relating to energy issues as a result of the process of industrialisation[7]. Rural, landscape issues were not, therefore – unlike in Britain and the United States – the main catalyst and Germany remains a key player in urban environmental history.

Where German Environmental History may, more recently, have developed a particular link with American Environmental History is in the relationship between environmentalism and Environmental History. Given Germany's recent history in relation to nuclear power in particular, there has certainly been a concern among some German environmental historians, such as Professor Radkau, to describe how we humans have got the planet, and ourselves, into the mess currently highlighted by the environmental movement[8].

Both of the above may help to explain why so many German environmental historians have preferred to work on the modern era (from the late 18th century through 1950s/70s). This period encompasses the Enlightenment through Improvement and Industrialisation to the eventual challenge to Enlightenment's central maxim that progress, both social and technological, was inevitably good. This last element plays a crucial role in modern environmentalism and also finds explicit expression in the work of the American environmental historian, Carolyn Merchant[9]. Nevertheless, as environmental histories of the modern period have become comparatively well-researched, environmental histories of the early modern and medieval periods are now becoming more popular in Germany, perhaps in part to provide a context for the modern period, but also on their own account[10].

6 For example *Rita Gudermann*, Morastwelt und Paradies. Ökonomie und Ökologie in der Landwirtschaft am Beispiel der Meliorationen in Westfalen und Brandenburg (1830-1880), Paderborn 2001; Naturkatastrophen. Beiträge zu ihrer Deutung, Wahrnehmung und Darstellung in Text und Bild von der Antike bis ins 20. Jahrhundert, hg. v. *Dieter Groh*, *Michael Kempe* und *Franz Mauelshagen*, Tübingen 2003, *Jens Ivo Engels*, Umweltgeschichte als Zeitgeschichte, in: Aus Politik und Zeitgeschichte 13 (2006), 32-38; Saeculum Theme Issue "Infrastrukturen und natürliche Ressourcen", to be published in 2007.

7 Blauer Himmel über der Ruhr. Geschichte der Umwelt im Ruhrgebiet 1840-1990, hg. v. *Franz-Josef Brüggemeier* und *Thomas Rommelspacher*, Essen 1992.

8 *Joachim Radkau*, Aufstieg und Krise der deutschen Atomwirtschaft 1945-1975. Verdrängte Alternativen in der Kerntechnik und der Ursprung der nuklearen Kontroverse, Reinbek 1983. On the history of the environmental movement in Germany cf. Kai Hünemörder's paper.

9 *Carolyn Merchant*, The Death of Nature. Women, Ecology, and the Scientific Revolution, San Francisco 1983.

10 For example *Martin Knoll*, Umwelt – Herrschaft – Gesellschaft. Die landesherrliche Jagd Kurbayerns im 18. Jahrhundert, St. Katharinen 2004.

Despite such positive activity, it must also be said that Environmental History in Germany is still in a poor state of institutionalization. There is no major review, nor learned society, nor academic association. Since the structure of German universities does not encourage the establishment of many specialized posts, Environmental History scarcely features as the explicit field of study among, for example, the holders of professorial chairs. Most environmental historians therefore hold posts in social, economic or urban history.

However, since 2004, a graduate school at the University of Göttingen has devoted itself to the "*interdisciplinary history of the environment*", while at Darmstadt Technical University, a centre of urban and Environmental History studies has recently been established. In addition, some informal centres housing important research teams devoted to Environmental History now exist in Cottbus, Bielefeld, Munich, and Freiburg. It is therefore perhaps not surprising that, despite (and in part, because of) the presence of important well-established individuals, the research field in German Environmental History is dominated by younger practitioners, often at PhD level, exploring the relationships between humankind and nature in very different ways. However, as Franz-Josef Brüggemeier states in his paper, the public visibility of Environmental History in Germany can still be improved.

In Britain, as indicated in Professor Smout's paper, natural history and historical geography were both extremely important to the development of what ultimately became Environmental History. Indeed, it has been argued that by "*arrogating to themselves a term already being used by at least two other disciplines (geologists and archaeologists), the historians managed to upset the self-esteem of a very particular group of scholars, the historical geographers*"[11]. While this is a fair point, there are grounds for hope that the historical geographer and the environmental historian now view each other as kindred spirits, rather than rivals. There is certainly much that the historian can learn from the geographer, in terms of problematising and systematising past landscapes, ecosystems and communities, both human and non-human. However, the degree of theoretical abstraction that has afflicted cultural geography in particular in the wake of postmodernism is fortunately not a route that historians have generally shown much interest in following.

This lineage back towards natural history and geography also gives Environmental History in Britain a degree of affinity with American Environmental History, which also emerged in part from geographical and ecological writings. This association has become easier, too, now that American environmental historians have been prepared to challenge the idea of their own pristine wildernesses, allowing comparisons to be made with the much more obviously 'lived in' landscapes of Europe. Urban environmental history, particularly for the post-industrial period, is also very much a feature of both American and British historiography (as it is in Germany), though it should be noted that, in both cases, rural environmental history is still the dominant strand.

[11] *Richard H. Grove*, Environmental History, in: New Perspectives on Historical Writing, Peter Birke ed., 2nd ed., Bodmin 2001, 261.

This is particularly interesting, from a cultural point of view, given that Great Britain's population is overwhelmingly urban. It should also be noted that many British urban environmental historians, especially in relation to pollution, are scientists or geographers by training.

In essence, it could be argued that British Environmental History, when practised by historians, is partly an attempt to catch up with other disciplines which are already busy understanding the role of humans in explaining environmental change. Historians are therefore only belatedly, and partially, stepping in to fulfil their traditional role in an area of study where others have already attempted, with varying degrees of success, to do their job for them. Where the British environmental historian is perhaps best able to play a distinctive role is in emphasising the role of the environment in influencing social, cultural, political and economic evolution within human communities.

However, in doing so, they risk incurring criticism from environmental historians from a scientific background, such as Oliver Rackham, for not understanding and integrating ecological conditions and responses, at an appropriate scale, with this human history[12]. Maintaining a balance of equality of understanding between these two complex systems – human and non-human – is, to be honest, beyond the capability of most individuals. It also limits exceptional individuals able to combine scientific and historical methodologies to very specific subject areas. As a result, it is beginning to be argued that the interdisciplinarity which Environmental Historians claim is intrinsic to their subject area should not be confined to attempts to integrate the secondary material of other disciplines to primary documentary research. Rather, it should encourage the working together of disciplines relevant to a particular research question, in order to generate first-class research across, and equality among, all its aspects. However, institutional mechanisms within academia in Britain do not yet make it very easy to accomplish such a goal. Equally, this relationship with the sciences and social sciences has also meant that Environmental History is not yet fully acknowledged as a part of mainstream history.

As with Germany, it could not yet be said that Environmental History in Britain is fully institutionalised. The AHRC Research Centre for Environmental History, situated at the Universities of Stirling and St Andrews in Scotland, is one of only a very few such Centres in Europe. However, it is dependent on funding from external sources and its future is very uncertain. Elsewhere, there are individual practitioners at the Universities of Nottingham and Newcastle, which, along with St Andrews, Stirling and Dundee, teach Environmental History at both undergraduate and postgraduate level. As with Germany, most environmental historians are scattered across UK universities in the guise of other sub-disciplines of history or even other disciplines. This last group include the historical geographers, such as Ian Simmons at Durham and Robert Dodghson at Aberystwyth, who have been pioneers in the field

[12] Dr Rackham's criticisms were made most explicitly to Fiona Watson, as Managing Editor of the Journal *Environment and History*, of which Dr Rackham was a member of the Editorial Board.

in Britain, as well as scientists like Oliver Rackham at Cambridge and Peter Brimblecombe at the University of East Anglia. Finally, the Centre for Documenting Environmental Change at the University of Cambridge is another important milieu through which Environmental History is fostered. Interestingly, its key player, Paul Warde, is actually a German historian.

As with any field of research seeking acceptance and, ultimately, permanence, Environmental History has made great progress in both Germany and Great Britain, but is still comparatively insecure, both within an institutional context and at a wider public level of awareness. Conferences such as this one can only help to promote a greater confidence in Environmental History's future in both countries.

Nevertheless, the papers presented at the Prince Albert Society's conference in September 2005 at Coburg and the discussions proved to be very fruitful and vivid, showing a historical discipline in full vigour.

Environmental Historiography

Christopher Smout traces back the origins of British Environmental History to the Enlightenment. He insists on the fact that Environmental History should be written as a combination of natural factors and human action, sketching a transdisciplinary approach which relies on evidence gathered by natural sciences such as palynology or the analysis of soils. The combination of the documentary historian's approach with the techniques of applied physical science and geography enables environmental historians to detect the impact of human activity on the natural environment over centuries and even, though unsupported by documentary material, millennia. Smout pays special attention to the work of Oliver Rackham, which has influenced public perception of woodlands and the landscape in general as historical artefacts[13].

Whereas the above-mentioned approaches have flourished since the 19th century (and still do), British Environmental History has expanded since around 1970, when historians began to be interested in the history of ideas and perceptions about nature. Since the 1990s, economic and social historians have explored an ever-growing range of environmental factors. However, it must be said that other branches of academic history do not pay enough attention to the findings of Environmental History.

Smout wonders why the American school of Environmental History has had relatively little impact on British landscape and agrarian history. According to him, the American conception of misunderstanding and violent misuse of nature did not fit with the British experience, where a relatively sustainable use of natural resources has prevailed for a long time, at least until comparatively recently.

With respect to Germany, by contrast, *Franz-Josef Brüggemeier* shows the important impact of both political ecology and American Environmental History on the early German historians committing themselves to looking at the environment. Ger-

[13] See particularly: Trees and Woodland in the British Landscape, London 1993.

man Environmental History is closely tied up with recent developments; Brüggemeier dates it back to the seventies and eighties. He points out that the highly normative approaches, which often aimed at interpreting human history in a totally new way, have lost their persuasiveness as much more empirical evidence has been collected. In short, the professionalization of Environmental History has led to more modest approaches. In Germany, Environmental History has been very close to the history of technology and to social, economic and urban history.

During the late eighties and nineties, two major issues dominated the debate. First, the history of wood scarcity at the end of the 18th century and the growing use of coal as an energy source attracted scholars' attention. Second, the history of industrial and urban pollution since the late 19th century has also been widely discussed. So, analyses of the energy sources underpinning the industrial era were, and are, a recurring theme in German Environmental History. More recently, however, Environmental History has expanded into a great variety of new fields such as the history of nature protection and environmentalism, natural disasters, and infrastructures. These findings will hopefully be increasingly integrated into historical debates *per se*.

Resource use and sustainability

The use of natural resources and the concept of sustainability play an important role on the agenda of environmental historians in both Britain and Germany. One of the most highly debated topics of environmental as well as economic history is the role of the commons, an issue that encompasses both legal and institutional aspects, as well as ecological impacts and processes.

Angus Winchester examines the patterns of resource use and management of common grazings in the uplands of northern England, which is still partly common land, from the 12th to the 20th century. He points out that the history of natural resources, as managed by humans, gives important insights into the structures of society, the patterns of power and cultural perceptions. In his paper he identifies two major transitions as a result of important social shifts, from seigniorial hunting ground to community grazing land in the late-medieval period, and from local resource to a resource open for (national) exploitation since the 19th century. Criticizing the still popular idea of the "tragedy of the commons", he insists on the fact that the use of common land in this context was always regulated. Consequently, there is no evidence that common land was overexploited. The exploitation of the uplands often seems to have originated in medieval times, as a result of the desire of a noble owner to raise income by allowing grazing. In the early modern period, two principles had already emerged to manage the access to these grazings in a spirit of "good neighbourhood": the rule of "levancy and couchancy" and "stint". The first principle stated that a tenant could not graze on the manorial wasteland in summer any more stock than he could keep during the winter on the produce of his own holding of land. The alternative was "stint", allowing access in terms of agreed numbers of livestock. By the time of enclosures and capitalist agriculture around 1800, these management systems collapsed and, finally, gave way to overexploitation. Winchester concludes that the early modern management of common grazings resulted in a sustainable use of resources, whereas medieval and modern exploitation have been far less successful.

The challenges of industrialisation and urbanisation

In political debates, as well as in Environmental History, industrialisation is interpreted as *the* major turning point in the relationship of humankind with its surrounding environment, enabling modern societies to manipulate natural resources to a hitherto unprecedented degree. While the polluting effects of big industries dominated debates in early Environmental History, recent research has explored the link between industrialisation and urbanisation. Two papers in this book deal with the environmental significance of urban development.

Bill Luckin addresses the debate on the rural and the urban since the mid-19th century. One of the characteristics of urbanism in industrial Britain was the assumption that the countryside had to be protected from the urban, due to an anti-urbanism and anti-metropolitanism deeply rooted in British culture and history. Whereas liberals in the early 19th century had greeted the urban centres as hotspots of freedom and progress, negative perceptions prevailed from the late 19th century on, depicting the cities as dangerous to health, happiness and morality. Most of the proposed solutions privileged a utopian move back to the country and can be interpreted as anti-modern and anti-industrial. The two World Wars occasioned a reinforcement of the image of rural England as a source for national regeneration in order to eradicate the frightening memories of human losses and destruction. After 1945, however, urban development was meant to *integrate* the principles of a good life which could be found in the countryside. Luckin underlines the fact that the anti-urban discourse did not take into account the social and economic reality of urban development. One of the most important developments since the mid-19th century was the massive process of suburbanization, highlighting the inadequacy of the perceived rural-urban dichotomy. Luckin pleads for further theorization of suburbanization, in order to integrate this variable into British urban history.

Dieter Schott underlines the interconnectedness of industrialisation, urbanisation and technical improvements, dramatically altering the human-nature relationship. He points out that industrialisation had at least two consequences. First, it caused new problems as a result of pollution. Second, it provided new solutions for environmental problems resulting from the urbanisation processes. However, every technical solution in turn brought new problems with it. Schott explains that environmental problems were openly discussed in late 19th century towns, more or less exclusively in terms of health and the economy/defence of ownership. British hygienic debates and solutions influenced political elites and technical experts in Germany. The management of water provides one of the most important examples. Water sewage systems constituted very complex technical infrastructures, altered the water cycle, improved living conditions in the cities, but resulted in heavy pollution loads in the big rivers. The second example cited in this paper is energy supply and the different strategies employed to reduce air pollution through town planning. Schott underlines the social aspects of all these solutions, which often favoured the better-off.

Looking after the environment

No overview of Environmental History is complete without a look at nature conservation and environmental protection. In both countries, the United Kingdom and Germany, the conservation movement has an important history. *Ian Simmons* calls attention to the fact that every society using natural resources has to protect them in one way or another. However, his paper concentrates on the industrial era, describing measures taken against pollution, caused by industrial production. From the mid-19th century onwards, British legislation sought to control industrial emissions. Since the early 20th century, the use of land became the second major object of protective policies, the planning Acts of the 1940s being one of its most important instruments. The history of National Parks after World War II and their governing body, the Nature Conservancy, highlights the importance of establishing a specialised public administration. More recently, the European Commission has taken on the role of an important administrative player in environmental protection. However, the outcomes of roughly one and a half centuries of environmental protection are ambivalent. The lack of enforcement and effects of competing interests thwarted many regulations. With respect to protection practices, Simmons concludes that the principle of selective protection still dominates, bringing into question the idea of an ecological revolution around 1970.

Kai Hünemörder focuses on the history of German nature conservation. He presents the major developments during the 20th century, privileging three major phases and the example of the province of Schleswig-Holstein. The paper elucidates the interrelatedness between regional, national and international developments, highlighting the fact that nature is only truly palpable at the local level. Around 1900, the middle class nature conservation movement emerged and developed a set of conservation schemes. The first laws, administrative protection measures and the founding of a nature conservation agency attest to the enormous political and cultural impact of this civic movement. The era of the Third Reich proved to be highly ambivalent in terms of nature conservation. On the one hand, the Nazi regime supported conservation by passing very modern nature protection laws and by restructuring the conservation bureaucracy. Moreover, it encouraged the integration of modern planning instruments in conservation schemes. On the other hand, World War II and the politics of autarky thwarted these innovations and caused appalling damage to the natural environment. Finally, many conservation officials were implicated in warcrimes as a result of the environmental planning they had imposed on the occupied territories in Eastern Europe. By 1970, the modern ecological era changed the face of conservation, transforming it from a national into an international affair. On the Schleswig-Holstein level, environmental consciousness enabled the establishment of large-scale protection covering the coastal zones.

Prince Albert Medals

Finally, *David Allan* gives an absorbing and wide-ranging exposé of the history of the Prince Albert Medal, which was first presented in 1863, two years after the Prince's untimely death, to Sir Rowland Hill, the founder of the modern postal service. What makes this account so fascinating within the context of this particular conference is the relationship between advances in science, arts and commerce, which the Royal Society of Arts was founded to promote, and environmental concerns. Within the modern context, much debate centres around the argument that progress, in economics, but most particularly in technology, is antithetical to a concern for the environment. However, the recent presentation of a Prince Albert Medal to Tim Smit, for services to ecology through his pioneering work on the Eden Project, clearly indicates that all these elements can – and, indeed, must – work together to provide a viable and sustainable future for both humankind and the rest of nature.

Conclusions

The above papers provide a useful summary of the level and focus of debate in Environmental History in both Germany and Great Britain. It should be clear that many of the subject's concerns fit extremely well with more mainstream historical debates, such as urbanisation and industrialisation, the legal and infrastructural underpinnings of resource use and landholding. We hope also to have made some comment on the potential role of History itself in interacting with other disciplines across the social sciences and even the sciences.

This conference has proved an extremely useful and fruitful historiographical exercise which has allowed some of Environmental History's practitioners to reflect on where the subject stands today in our two countries. Such an exercise is essential, from time to time, as an effective means of exposing the gaps and weaknesses, as well as the successes, and highlighting the way forward. Above all, it has shown that historians have an important contribution to make in understanding the long-term perspective on the complex human relationship with non-human nature, a perspective that is essential, among others, for the future of that relationship. There is no doubt about the immense political and social importance of environmental issues which will continue to occupy national and international politics in the following decades.

Fiona Watson – Jens Ivo Engels

Einleitung

Als die Prinz-Albert-Gesellschaft im September 2005 eine Tagung zum Stand der Umweltgeschichte in Großbritannien und Deutschland ausrichtete, so geschah dies im Bewusstsein des großen Interesses, das der Prinzgemahl Albert für die Landwirtschaft zeigte, und des enormen Einflusses, den er zusammen mit seiner Frau auf die Popularisierung der schottischen Highlands ausübte, die zu den wildesten und beeindruckendsten Landschaften des Vereinigten Königreichs zählen. Zudem lag es den Organisatoren am Herzen, diesen relativ neuen Forschungszweig den Mitgliedern der Prinz-Albert-Gesellschaft und Vertretern anderer historischer Disziplinen vorzustellen.

Ziele und Probleme der Umweltgeschichte

Die wichtigste Aufgabe der Umweltgeschichte besteht darin, der Natur als einem historischen Faktor neben all den anderen Forschungsinteressen in der Geschichtswissenschaft Geltung zu verschaffen. Sie versucht zu verstehen, wie die Umwelt die menschlichen Lebensbedingungen formte und beeinflusste, und umgekehrt, wie der Mensch sich auf diese Bedingungen einstellte und die natürliche Umwelt veränderte. Anders als noch vor zwei Jahrzehnten beanspruchen Umwelthistoriker heute nicht mehr, die Menschheitsgeschichte neu zu schreiben; auch bleibt die reine Naturgeschichte (ohne menschlichen Einfluss) in der Regel ausgeblendet. Vielmehr beschreibt die Umwelthistorie die Beziehungsgeschichte zwischen Menschen und der nicht-menschlichen Natur als Teil der Geschichte *unserer* Spezies.

Wie alle anderen historischen Disziplinen spiegelt die Umweltgeschichte die Problemstellungen in der Epoche ihrer Protagonisten wider. Es fällt auf, dass ein Zweig der Umweltgeschichte sich in einigen Ländern, insbesondere des Südens, aus der Kolonial- bzw. Imperialgeschichte entwickelte, wobei die Vorstellung des „clash of civilisations" mit dem eines „clash of species" gekoppelt war. Klassischerweise weist die Umweltgeschichte drei wesentliche Untersuchungsperspektiven auf[1]. Die erste besteht darin, die wichtigsten Trends in der Veränderung der Umwelt über längere Zeit-

[1] *J.R. McNeill*, Observations on the Nature and Culture of Environmental History, in: History and Theory, Theme Issue 42 (2003), 5-43.

räume zu untersuchen, so wie es die Klimageschichte und die Geschichte der Bodennutzung tun. Zum zweiten untersucht die Umweltgeschichte jene Strategien des Menschen, die dem Umgang mit diesen Veränderungen gewidmet waren. Diese Herangehensweise kann auch als „institutionelle“ Umweltgeschichte bezeichnet werden, da sie häufig legislative oder verwaltungsmäßige Praktiken in den Blick nimmt – sei es von Seiten des Staates oder privater Organisationen. Drittens haben Naturvorstellungen und deren kulturelle Bedingtheit die Aufmerksamkeit vieler Historiker auf sich gezogen. Dabei entstand eine große Bandbreite von Jahrhunderte übergreifenden Überblicksstudien bis hin zu Arbeiten, die sich der Wahrnehmung einzelner Landschaften oder Arten widmen. In letzter Zeit fand die Geschichte der Mensch-Tier-Beziehungen zunehmendes Interesse, ebenso wie das Problem sporadischer oder serieller Mensch-Natur-Konfrontation (vor allem in der Katastrophengeschichte).

Da Umwelt und menschliche Geschichte auf komplexe Weise miteinander verflochten sind, überrascht es nicht, wie vielfältig die Zweige der Umweltgeschichte mittlerweile sind; eine allumfassende Geschichte der Mensch-Natur-Beziehungen ist folglich kaum möglich. Zunächst schien es so, als könne die Ökologie ein integrierendes Modell zu ihrer Beschreibung liefern, doch versagen diese Ansätze insbesondere dann, wenn es um die Analyse der menschlichen Gesellschaft geht; deren Komplexität erfasst die ökologische Modellbildung nicht. Ein weiterer Ansatz – das Konzept der historischen Energie-Regime, welches den ökologischen Fußabdruck der Gesellschaft beschreiben möchte – ist in ähnlicher Weise dafür kritisiert worden, die menschlichen Systeme zu vereinfacht darzustellen. Tatsächlich haben die Umwelthistoriker bislang keine Methode entwickelt, die es ihnen erlauben würde, die Mensch-Natur-Beziehungen in ihrer Vielfalt kohärent abzubilden. Dennoch bleibt zu hoffen, dass mittelfristig ein methodischer Rahmen gefunden wird, der einerseits die fachinternen Befunde miteinander kompatibel macht, andererseits aber der Komplexität menschlicher Geschichte angemessen Rechnung trägt.

Schließlich müssen die Umwelthistoriker ähnlich wie Philosophen mit einem weiteren Problem leben, das Joachim Radkau auf den Punkt gebracht hat[2]. Einerseits ist die uns umgebende Natur in starkem Maß durch menschliche Aktivität beeinflusst. Insofern ist etwa die Landschaft ein Artefakt, das aus wissenschaftlicher, technischer, kultureller und philosophischer Perspektive betrachtet werden kann. Andererseits existiert die Natur aber auch jenseits jeglichen Handelns, ja ist oftmals sogar menschlicher Erfahrung entzogen. Dieser Befund bringt erkenntnistheoretische Paradoxien mit sich. Ähnlich wie es für die Quantenphysik am Beispiel von Schrödingers Katze gezeigt wurde, ist es unmöglich, diese außerhalb der menschlichen Sphäre existierende Natur zu „messen“ oder zu „beschreiben“, denn jeder derartige Versuch führt zur Veränderung des Untersuchungsobjekts und macht es damit ein Stück weit zum künstlichen Objekt.

[2] *Joachim Radkau*, Was ist Umweltgeschichte?, in: Umweltgeschichte, hg. v. Werner Abelshauser, Göttingen 1994, 11-28.

Zum Stand der Umweltgeschichte in Großbritannien und Deutschland

Wie bereits erwähnt, zeichnet sich die Umweltgeschichte durch ein breites Spektrum an Zugangsweisen und Themenstellungen aus. Wenn auch in letzter Zeit der Versuch gemacht wurde, auf der Grundlage von umwelthistorischen „Wendepunkten" einen Referenzrahmen für das Fach zu erstellen, so zeigen sich bei solchen Anlässen stets die großen Differenzen, welche sich oft aus den unterschiedlichen geokulturellen Perspektiven der Forscherinnen und Forscher ergeben[3]. In den Vereinigten Staaten, ähnlich wie in Australien, begann die Entfaltung der modernen Umweltgeschichte vor etwa dreißig Jahren. Von Beginn an stand das Thema „Wildnis" im Mittelpunkt der Debatte, vermutlich weil die riesigen, vielfach (scheinbar) unberührten Landschaften einen festen Platz in der nationalen Kultur dieser Länder haben. Zudem gibt es enge Beziehungen zwischen der Umweltgeschichte und der Umweltbewegung in Amerika, insbesondere bei ihren frühen Vertretern wie Donald Worster – eine Beziehung, die in den meisten anderen Ländern schwächer ausgeprägt ist.

Erst etwa eine Dekade später interessierten sich die Historiker in Europa für die Umwelt. In Großbritannien nicht zuletzt auch deshalb, weil ihre Bedeutung für die Geschichte des Empire und der Dritten Welt erkannt wurde. Hieraus entwickelte sich ein Bewusstsein für den Einfluss, den die Erfahrungen der Kolonisatoren beim Leben und Arbeiten in unterschiedlichen Klimazonen und Ökosystemen auf das intellektuelle Leben und die wissenschaftliche Forschung im Heimatland ausübten[4]. In diesem Kontext ist es beispielsweise möglich, den Einfluss der britischen Kolonien auf das Mutterland im Bereich der Forstwirtschaft nachzuzeichnen. Forstwissenschaftliche Zugangsweisen aus Deutschland wurden zunächst für die Waldwirtschaft in Indien adaptiert, anschließend wurden die dort gemachten Erfahrungen auf die Britischen Inseln übertragen – mit bisweilen bizarren Folgen. So wendete man in Schottland Techniken an, die ursprünglich dazu bestimmt waren, Bodenaustrocknung zu verhindern, obwohl Schottland bekanntermaßen nicht an Niederschlagsmangel leidet.

Zugleich entstand auch eine Geschichtsschreibung über die Umwelt des europäischen Kontinents, eine Entwicklung, die von der Gründung der Europäischen Gesellschaft für Umweltgeschichte – European Society for Environmental History – im Jahre 1999 erheblichen Rückenwind erhielt. Die Veranstaltungen der ESEH haben den ohnehin ausgeprägt internationalen Charakter der Umweltgeschichte noch verstärkt[5].

[3] Vgl. dazu die Ergebnisse einer Tagung in Bielefeld, die unter folgendem Titel publiziert werden: The Turning Points of Environmental History, hg. v. *Joachim Radkau* und *Frank Uekötter*, Lanham, im Druck.

[4] *Alfred W. Crosby*, Ecological Imperialism: the Biological Expansion of Europe, 900-1900, Cambridge 1986; Conservation in Africa, People, Policies and Practice, D. Anderson and Richard Grove eds., Cambridge 1987; *John M. MacKenzie*, Imperialism and the Natural World, Manchester 1990; *Richard H. Grove*, Green Imperialism: Colonial Expansion, Tropical Island Edens and the Origins of Environmentalism, 1600-1800, Cambridge and New Delhi 1995.

[5] www.eseh.org.

Trotz der Vielzahl umwelthistorisch arbeitender Forscherinnen und Forscher in ganz Europa ist die Umweltgeschichte auf unserem Kontinent in Deutschland und Großbritannien am ausgeprägtesten und einflussreichsten. Dennoch, so zeigen es die Beiträge von Christopher Smout und Franz-Josef Brüggemeier, sind die Wurzeln und die gewählten Zugänge in beiden Ländern fast so unterschiedlich wie im Vergleich mit den USA.

Die deutsche Umweltgeschichte entwickelte sich aus anderen Teildisziplinen der Geschichtswissenschaft – anders als in Großbritannien, wo der Einfluss der Naturwissenschaften und der Sozialwissenschaften sehr viel größer war. Folglich sind die Methoden und die Quellen der deutschen Umweltgeschichte denen der anderen historischen Zweige sehr ähnlich. Und dennoch entwickelte sich die deutsche Umweltgeschichte zunächst als eine spezialisierte Subdisziplin mit geringer Außenwirkung. In letzter Zeit haben sich die Umwelthistoriker bemüht, ihre Forschungen an die Debatten der Allgemeinen Geschichte anzuschließen. Dabei wurden neue Fragestellungen erprobt, so wie die Geschichte der Infrastrukturen, der Naturkatastrophen, Agrargeschichte, Wissenschaftsgeschichte, Politik- und Kulturgeschichte, Geschichte von Herrschaftsausübung etc.[6] In der Folge ist die Beachtung der Umweltgeschichte im historischen „Mainstream" gewachsen, wobei diese freilich noch ihre Grenzen hat. Die in diesem Buch versammelten Beiträge zeigen dies sehr deutlich, denn sie enthalten eine Fülle von Berührungspunkten mit und Erkenntnismöglichkeiten für andere historische Fragestellungen. So zeigt sich, dass die Ressourcennutzung immer eine Frage von Machtverhältnissen ist; umgekehrt war die urbane Entwicklung in den Städten des 19. und 20. Jahrhunderts stark von Umweltfaktoren beeinflusst; schließlich bietet die Geschichte des politischen Umweltschutzes wichtige Einblicke in die Funktionsweise politischer Systeme.

Die deutschen Historiker begannen vor etwa zwanzig Jahren, sich im Kontext der Geschichte der Industrialisierung mit Energiefragen und anschließend für die Umwelt zu interessieren[7]. Anders als in Großbritannien und den USA waren Fragen der Landschaftsentwicklung und der agrarischen Nutzung zunächst weniger wichtig.

Eine Ähnlichkeit zwischen deutscher und amerikanischer Umweltgeschichte besteht allerdings in dem großen Interesse für die Umweltbewegung. Viele deutsche Umwelthistoriker, allen voran Joachim Radkau, haben sich angesichts der in ihrem

[6] Z.B. *Rita Gudermann*, Morastwelt und Paradies. Ökonomie und Ökologie in der Landwirtschaft am Beispiel der Meliorationen in Westfalen und Brandenburg (1830-1880), Paderborn 2001; Naturkatastrophen. Beiträge zu ihrer Deutung, Wahrnehmung und Darstellung in Text und Bild von der Antike bis ins 20. Jahrhundert, hg. v. *Dieter Groh*, *Michael Kempe* und *Franz Mauelshagen*, Tübingen 2003; *Jens Ivo Engels*, Umweltgeschichte als Zeitgeschichte, in: Aus Politik und Zeitgeschichte 13/2006, 32-38; Saeculum Themenheft „Infrastrukturen und natürliche Ressourcen", erscheint 2007.

[7] Blauer Himmel über der Ruhr. Geschichte der Umwelt im Ruhrgebiet 1840-1990, hg. v. *Franz-Josef Brüggemeier* und *Thomas Rommelspacher*, Essen 1992.

Land heftigen Proteste etwa gegen die Kernenergie mit der Frage auseinandergesetzt, in welcher Weise der Mensch seine Umwelt schädigte[8].

Diese beiden Gesichtspunkte erklären, warum die meisten deutschen Umwelthistoriker ihre Forschungen in der Neueren Geschichte (vom späten 18. Jahrhundert bis zu den siebziger Jahren des 20. Jahrhunderts) verorten. Dieser Zeitraum umfasst die Aufklärung, den Aufstieg der Industriegesellschaft, schließlich aber auch die Infragestellung eines seit dem 18. Jahrhundert dominanten Fortschrittsdenkens, das die Kehrseiten der sozialen und vor allem technisch-wirtschaftlichen Modernisierung nicht erfasste. Die Abkehr von einer positiven Bewertung des Fortschrittsdenkens ist eine zentrale Prämisse der modernen Umweltbewegung und dominiert zum Beispiel das Werk der amerikanischen Umwelthistorikerin Carolyn Merchant[9]. In den letzten Jahren ist freilich ein zunehmendes Interesse der deutschen Umweltgeschichte für die Frühe Neuzeit und das Mittelalter festzustellen, und zwar sowohl als Vorgeschichte der immer noch dominierenden Umweltgeschichte in der Moderne, als auch mit ganz eigenen, epochengebundenen Fragestellungen[10].

Trotz zahlreicher Studien bleibt der Institutionalisierungsgrad der Umweltgeschichte in Deutschland schwach. Es gibt keine umwelthistorische Fachzeitschrift und auf nationaler Ebene keine Vereinigung oder akademische Gesellschaft (anders als für die Technik-, Wirtschafts-, Frauen-, oder auch britische Geschichte). Da deutsche Universitäten im Vergleich zu ihren britischen Gegenparts zurückhaltend bei der Schaffung spezialisierter Stellen sind, gibt es so gut wie keine der Umweltgeschichte gewidmeten Lehrstühle. Die meisten Umwelthistoriker haben daher Positionen, die offiziell der Sozial-, Wirtschafts-, Technik- oder Stadtgeschichte gewidmet sind.

Dennoch gibt es erste Anzeichen für eine institutionelle Verankerung der umwelthistorischen Forschung. Seit 2004 arbeitet in Göttingen ein Graduiertenkolleg zur *„Interdisziplinären Umweltgeschichte"* und an der TU Darmstadt entstand jüngst ein Zentrum für Stadt- und Umweltgeschichte. Daneben gibt es eine Reihe eher informeller Zentren mit einflussreichen umwelthistorischen Forschergruppen in Cottbus, Bielefeld, München und Freiburg. Es kann daher nicht überraschen, dass die deutsche Umweltgeschichte trotz einiger ausgewiesener und etablierter Forscherpersönlichkeiten hauptsächlich von Angehörigen des wissenschaftlichen Nachwuchses auf Doktorandenebene praktiziert wird. Jedenfalls ist die öffentliche und fachinterne Präsenz der Umwelthistorie in Deutschland verbesserungswürdig, wie Franz-Josef Brüggemeier in seinem Beitrag unterstreicht.

[8] *Joachim Radkau*, Aufstieg und Krise der deutschen Atomwirtschaft 1945-1975. Verdrängte Alternativen in der Kerntechnik und der Ursprung der nuklearen Kontroverse, Reinbek 1983. Zur Geschichte der Umweltbewegung in Deutschland vgl. den Beitrag von Kai Hünemörder in diesem Band.

[9] *Carolyn Merchant*, The Death of Nature. Women, Ecology, and the Scientific Revolution, San Francisco 1983.

[10] Beispielsweise: *Martin Knoll*, Umwelt – Herrschaft – Gesellschaft. Die landesherrliche Jagd Kurbayerns im 18. Jahrhundert, St. Katharinen 2004.

Christopher Smout betont in seinem Artikel die Bedeutung von Naturgeschichte und historischer Geographie für die Entwicklung dessen, was später Umweltgeschichte genannt wurde. Richard Grove hat es einmal so formuliert: *„Indem sie sich eine Bezeichnung anmaßten, die schon lange von mindestens zwei Disziplinen besetzt worden war (Geologen und Archäologen), gelang es den Historikern, das Selbstwertgefühl einer speziellen Gruppe von Forschern zu untergraben, nämlich das der historischen Geographen“*[11]. Auch wenn diese Bemerkung berechtigt ist, gibt es Grund zu der Hoffnung, dass Historische Geographie und Umweltgeschichte sich mittlerweile als verwandte Fachgebiete betrachten und nicht mehr als Rivalen. Tatsächlich spricht einiges dafür, dass Historiker von Geographen eine Menge lernen können, insbesondere in Hinblick auf die systematische Beschreibung historischer Landschaften, Ökosysteme und Lebensgemeinschaften von Menschen und Tieren bzw. Pflanzen. Glücklicherweise sind die meisten Historiker allerdings nicht dem Trend übertriebener theoretischer Abstraktion gefolgt, den die Kulturgeographie im Kontext postmoderner Theoriedebatten gezeigt hatte.

Die traditionelle Verbindung mit Naturgeschichte und Geographie verleiht der britischen Umweltgeschichte eine gewisse methodische Nähe zur amerikanischen, zumal auch hier einige Arbeiten von Geographie und Ökologie beeinflusst wurden. Seitdem die amerikanischen Kollegen begonnen haben, ihre Vorstellung von einer angeblich unberührten „Wildnis“ in Teilen ihres Kontinents zu revidieren und damit auch Vergleiche zu den noch deutlicher „bewohnten“ Landschaften Europas zulassen, ist die Zusammenarbeit einfacher geworden. Auch die städtische Umweltgeschichte erfährt in Großbritannien und den USA, ähnlich wie in Deutschland, viel Aufmerksamkeit, wenn auch zu betonen ist, dass die angelsächsische Forschungslandschaft nach wie vor von Fragestellungen mit Bezug zum ländlichen Raum geprägt ist. Das ist auch deshalb bemerkenswert, weil die britische Bevölkerung schon seit langem mehrheitlich in Städten wohnt. Schließlich sollte beachtet werden, dass auch ein großer Teil der mit Umweltfragen befassten britischen Stadthistoriker eine naturwissenschaftliche oder geographische Ausbildung besitzen.

Angesichts dieser Befunde kann behauptet werden, dass die Umweltgeschichte in Großbritannien aus Sicht der historischen Zunft ein Feld mit Nachholbedarf ist; andere Disziplinen haben dagegen bereits viel zum Verständnis des menschlichen Einflusses auf Umweltveränderungen beigetragen. Historiker im engeren Sinne sind eher verspätet und nur teilweise dabei, in diesem Feld ihre angestammte Rolle auszufüllen, während ihre Arbeit hier bisher von anderen gemacht wurde – mit unterschiedlichem Erfolg. Angesichts dieser Vorarbeiten wird die Aufgabe der ausgebildeten Historiker vor allem darin bestehen, die Rückwirkungen der Umweltbedingungen auf die sozialen, kulturellen, politischen und ökonomischen Verhältnisse in der Gesellschaft zu betonen. Dabei riskieren sie allerdings die Kritik der Umwelthistoriker mit naturwissenschaftlichem Hintergrund wie Oliver Rackham mit dem Vor-

[11] Das übersetzte Zitat stammt aus *Richard H. Grove*, Environmental History, in: New Perspectives on Historical Writing, Peter Birke ed., 2. Aufl., Bodmin 2001, 261.

wurf, die ökologischen Bedingungen und Interaktionen in ihrer am Menschen orientierten Geschichtsschreibung nicht zu verstehen oder auszublenden[12].

Allerdings ist es bei ehrlicher Betrachtung nur den wenigsten Forschern möglich, sowohl die komplexen Gesetzmäßigkeiten der natürlichen Systeme als auch die nicht minder komplexen Funktionsweisen menschlicher Gesellschaften in gleicher Weise zu verstehen. Selbst die Wenigen, denen es tatsächlich gelingt, naturwissenschaftliche und historische Methoden miteinander zu kombinieren, müssen sich auf eng begrenzte Themenfelder beschränken. Aufgrund dieser Erfahrung wird in letzter Zeit über die Arbeitsbedingungen der Interdisziplinarität nachgedacht, die nach Ansicht der meisten britischen Umwelthistoriker zum Kernverständnis ihres Faches gehört. Sie sollte weniger darin bestehen, Sekundärliteratur anderer Fächer mit der eigenen Analyse von Primärquellen zu kombinieren. Vielmehr sollten die Forschungsprojekte künftig in der Zusammenarbeit von Vertretern der unterschiedlichen Disziplinen entstehen, die sich einer gemeinsamen Forschungsfrage widmen, so dass die unterschiedlichen Herangehensweisen gleichwertige und gleichberechtigte Antworten liefern. Allerdings machen es die institutionellen Mechanismen innerhalb der britischen Forschungslandschaft nicht gerade einfach, ein solches Ziel zu verwirklichen. Zudem hat die enge Beziehung der Umweltgeschichte zur Naturwissenschaft auch dazu geführt, dass sie innerhalb der traditionellen Geschichtswissenschaft noch nicht vollkommen akzeptiert wird.

Ähnlich wie für Deutschland kann auch mit Blick auf Großbritannien noch nicht gesagt werden, dass die Umweltgeschichte voll institutionalisiert wäre. Das Forschungszentrum für Umweltgeschichte (*AHCR Research Centre for Environmental History*), das den Universitäten von Stirling und St. Andrews in Schottland angegliedert ist, stellt eines der sehr wenigen Zentren dieser Art in Europa dar. Seine Existenz ist jedoch von Drittmitteln abhängig und seine Zukunft daher nicht gesichert. Desweiteren gibt es einzelne Wissenschaftler an den Unversitäten von Nottingham und Newcastle. Hier gibt es wie in St. Andrews, Stirling und Dundee ein Lehrangebot in Umweltgeschichte, sowohl im Bereich undergraduate als auch postgraduate. Weitere Umwelthistoriker sind anderen historischen Subdisziplinen oder anderen Fächern zugeordnet. Zu letzteren gehören historisch arbeitende Geographen, wie die beiden Pioniere der Umweltgeschichte Ian Simmons in Durham und Robert Dodghson in Aberystwyth, und Naturwissenschaftler wie Oliver Rackham in Cambridge und Peter Brimblecombe an der University of East Anglia. Schließlich ist das Zentrum für die Dokumentation von Umweltveränderungen an der University of Cambridge eine wichtige Adresse. Interessanterweise spielt hier mit Paul Warde ein deutscher Historiker die zentrale Rolle.

Wie andere innovative Ansätze auf der Suche nach fachlicher Anerkennung hat die Umweltgeschichte in Deutschland und Großbritannien große Fortschritte gemacht,

[12] Rackham äußerte diese Kritik gegenüber Fiona Watson in ihrer Funktion als Geschäftsführende Herausgeberin der Zeitschrift *Environment and History*, in deren Herausgebergremium er Mitglied war.

ohne bislang eine vollkommen gesicherte institutionelle Basis und gleichbleibende Aufmerksamkeit in der Fachöffentlichkeit zu erhalten. Daher sind Veranstaltungen wie die von der Prinz-Albert-Gesellschaft im September 2005 in Coburg ausgerichtete Tagung wichtig, um für Interesse und Vertrauen in die Zukunft des Faches in beiden Ländern zu werben. Tatsächlich zeigten die facettenreichen Vorträge und lebhaften Diskussionen ein historisches Teilfach voller Vitalität.

Umweltgeschichtsschreibung

Christopher Smout schildert die Ursprünge der britischen Umweltgeschichte seit der Aufklärung. Er unterstreicht, dass sie stets als eine Kombination von natürlichen Faktoren und menschlichem Handeln geschildert werden sollte. Dabei skizziert er einen transdisziplinären Ansatz, der auch auf die Ergebnisse naturwissenschaftlicher Forschung zurückgreift, etwa aus der Pollen- und der Bodenanalyse. Erst die Kombination aus den Ansätzen des mit schriftlichen Quellen arbeitenden Historikers und den Techniken der angewandten physischen Wissenschaften, sowie der Geographie, versetzt den Historiker in den Stand, den Einfluss menschlicher Aktivitäten auf die Umwelt über Jahrhunderte und sogar Jahrtausende verfolgen zu können, wenn auch im letzten Fall ohne die Grundlage schriftlicher Quellen. Smout erörtert insbesondere die Arbeiten von Oliver Rackham, der ein öffentliches Bewusstsein dafür gefördert hat, dass Wälder und Landschaften historische Artefakte sind[13].

Während die bereits genannten Ansätze sich seit dem 19. Jahrhundert bis heute entwickelt haben, weitete sich die britische Umweltgeschichte seit etwa 1970 aus, als Historiker begannen, sich für die Geschichte von Naturideen und Naturwahrnehmung zu interessieren. Seit den neunziger Jahren schließlich untersuchen auch Wirtschafts- und Sozialhistoriker eine ständig wachsende Zahl von Umwelteinflüssen in der Geschichte. Leider, so bleibt festzustellen, beachten andere Zweige der Geschichtswissenschaft die Ergebnisse der Umweltgeschichte nicht in ausreichender Weise.

Smout stellt zudem die Frage, warum die US-amerikanische Schule der Umweltgeschichte verhältnismäßig geringen Einfluss auf die britische Landschafts- und Agrargeschichtsforschung ausübte. Seiner Ansicht nach ist einer der Hauptgründe darin zu sehen, dass die den meisten amerikanischen Arbeiten zugrundeliegende Vorstellung eines fehlenden Naturverständnisses und brutaler Ausbeutung sich nicht mit den britischen Erfahrungswerten deckt. Bis vor kurzem herrschte auf der Insel nämlich ein relativ nachhaltiger Umgang mit natürlichen Ressourcen.

Mit Blick auf Deutschland betont *Franz-Josef Brüggemeier* dagegen den großen Einfluss der politischen Ökologie und der amerikanischen Umweltgeschichte auf diejenigen Historiker, die sich als erste mit Umweltthemen befassten. Die Entstehung der deutschen Umweltgeschichte in den siebziger Jahren ist nach seinen Worten

[13] Vgl. insbesondere Trees and Woodland in the British Landscape, London 1993.

eng mit diesen Entwicklungen verbunden. Allerdings unterstreicht er auch, dass die stark normativen Ansätze der Frühzeit, die für sich in Anspruch nahmen, die Menschheitsgeschichte aus dem Blickwinkel der Umwelt vollkommen neu zu erzählen, ihre Überzeugungskraft in dem Maße verloren, wie empirische Untersuchungen zunahmen. Kurz gesagt, bewirkte die voranschreitende Professionalisierung der Umweltgeschichte eine Hinwendung zu bescheideneren Zugängen. In Deutschland dominierte bald eine enge Verbindung zwischen Umwelthistorie und Technik-, Wirtschafts-, Sozial- sowie Stadtgeschichte.

In den späten achtziger und in den neunziger Jahren beherrschten vor allem zwei Themenkomplexe die Fachdebatte. Zum einen beschäftigten sich die Forscher mit der Geschichte der Holzknappheit am Ende des 18. Jahrhunderts und der späteren Umstellung auf Kohle als Energielieferanten. Zum zweiten wurde intensiv über industrielle und städtische Umweltverschmutzung seit dem späten 19. Jahrhundert diskutiert. Die Beschäftigung mit den energetischen Grundlagen des Industriezeitalters war und ist weiterhin einer der Schwerpunkte der deutschen Umweltgeschichte. In letzter Zeit weiteten sich die umwelthistorischen Ansätze jedoch in erheblicher Weise aus. Zu den neuen Untersuchungsfeldern gehören die Geschichte des Natur- und Umweltschutzes, von Naturkatastrophen und Infrastrukturen. Die Ergebnisse dieser Forschungen werden hoffentlich zunehmend in die historischen Fachdebatten integriert.

Ressourcen und Nachhaltigkeit

Ressourcennutzung und das Konzept der Nachhaltigkeit spielen sowohl in der deutschen wie auch in der britischen Umweltgeschichte eine wichtige Rolle. Ein Thema, das sowohl in der Umwelt- als auch in der klassischen Wirtschaftsgeschichte breiten Raum einnimmt, stellen die britischen „commons" dar, gemeinschaftlich genutzte Agrarflächen. Das Beispiel der commons umfasst wichtige juristische und institutionelle Aspekte ebenso wie ökologische Auswirkungen und natürliche Prozesse.

Angus Winchester untersucht in seinem Beitrag die Nutzung und Pflege von Weideland in Nordengland, das teilweise bis heute in gemeinschaftlichem Besitz ist, und zwar für den langen Zeitraum vom 12. bis zum 20. Jahrhundert. Er hebt hervor, dass die Geschichte natürlicher Ressourcen bzw. die Pflege der Ressourcen wichtige Einblicke in die jeweiligen Gesellschaftsstrukturen vermittelt, insbesondere mit Blick auf Machtverhältnisse und kulturell vermittelte Wahrnehmungen. In seinem Beitrag identifiziert er zwei epochale Veränderungsschübe, die er auf soziale Wandlungen zurückführt, nämlich die spätmittelalterliche Transformation von grundherrlichen Jagdrevieren in gemeindliches Weideland und die Verwandlung einer lokal gepflegten Ressource in eine Landschaft, die prinzipiell der (landesweiten) Ausbeutung offen stand, ein Prozess, der sich seit dem 19. Jahrhundert beobachten lässt. Winchester kritisiert die immer noch verbreitete These von der *„Tragödie der Allmende"*; stattdessen belegt er ausführlich, dass der Zugang zum gemeinschaftlich genutzten Land stets streng reguliert war. Folglich gibt es auch keine Belege, dass es übernutzt wurde. Die landwirtschaftliche Inanspruchnahme der Uplands geht in den meisten Fällen vermutlich auf das Spätmittelalter zurück, als adelige Eigentümer ihr Einkommen steigerten, indem sie ihr Land zur Beweidung zur Verfügung stellten. In der Frühen Neuzeit hatten sich zwei Methoden herausgeschält, um den Zugang zu den Weiden nach dem Prinzip der „guten Nachbarschaft" zu regulieren, nämlich die Regel von „levancy and couchancy" und der „stint". Die erste Regel besagte, dass niemand während des Sommers mehr Vieh auf die Weide schicken durfte, als er im Winter auf der Grundlage der Erträge seines eigenen Landes versorgen konnte. Die andere Methode, der „stint", legte eine exakte Anzahl von Tieren fest, die Zugang zur Weide erhalten durften. Um 1800, als die massiven Einhegungen durchgeführt waren und die kapitalistische Agrarwirtschaft dominierte, brachen die frühneuzeitlichen Regelungssysteme zusammen, was in der Konsequenz zur Übernutzung des Weidelandes führte. So schließt Winchester mit der Feststellung, dass die frühmoderne Regulierung der Weide eine nachhaltige Ressourcennutzung darstellte, während mittelalterliche und moderne Nutzungsformen langfristig weniger erfolgreich waren bzw. sind.

Die Herausforderungen von Industrialisierung und Urbanisierung

Die Industrialisierung gilt in politischen Debatten ebenso wie in der Umweltgeschichte als *der* entscheidende Wendepunkt in der Beziehung von Mensch und Umwelt. Sie ermöglichte es den modernen Gesellschaften, natürliche Ressourcen in bislang ungekannter Weise zu nutzen und zu verändern. Während die Verschmutzung durch die Großindustrie die Debatten in der früheren Umweltgeschichte dominierte, wandte sich die Forschung in jüngerer Zeit dem Zusammenhang von Industrialisierung und Urbanisierung zu. Zwei Beiträge in diesem Buch sind daher umweltbezogenen Aspekten urbaner Entwicklung gewidmet.

Bill Luckin beschäftigt sich mit der Debatte über den Gegensatz von Stadt und Land seit der Mitte des 19. Jahrhunderts. Eine der Grundannahmen im britischen Urbanismus des Industriezeitalters war die Vorstellung, das Land müsse vor dem Einfluss der Stadt geschützt werden. Diese Vorstellung speiste sich aus einer Stadtfeindschaft und insbesondere Hauptstadtfeindschaft, die tief in der britischen Kultur und Geschichte verwurzelt sind. Während liberale Denker im frühen 19. Jahrhundert die urbanen Zentren noch als Hort von Freiheit und Fortschritt gefeiert hatten, gewannen skeptische Stimmen seit dem späten 19. Jahrhundert die Oberhand und bezeichneten das Großstadtleben als Gefahr für Gesundheit, Wohlbefinden und Moral. Die meisten Lösungsvorschläge orientierten sich an einer utopischen Rückkehr auf das Land, waren antimodern und zugleich industriefeindlich. Die beiden Weltkriege verstärkten die Orientierung am Bild eines ländlichen England als Quelle nationaler Regeneration und Möglichkeit, die beklemmende Erinnerung an menschliche Verluste und Zerstörungen vergessen zu machen.

Dennoch vollzog sich nach 1945 insofern ein Wandel, als die Stadtentwicklung nunmehr darauf abzielte, jene Prinzipien des guten Lebens im städtischen Raum zu verwirklichen, die angeblich nur auf dem Lande zu finden waren. Luckin unterstreicht indes, dass der stadtfeindliche Diskurs die sozialen und wirtschaftlichen Realitäten der urbanen Entwicklung nicht zur Kenntnis nahm. Eines der wichtigsten Phänomene war seit Mitte des 19. Jahrhunderts der Prozess der Suburbanisierung, der deutlich zeigt, wie wenig die angebliche Dichotomie von Stadt und Land den realen Verhältnissen entsprach. Luckin plädiert abschließend dafür, das Phänomen Suburbanisierung konzeptionell exakt zu erfassen, um es als bedeutenden Faktor der britischen Stadtgeschichte angemessen würdigen zu können.

Dieter Schott unterstreicht in seinem Beitrag die Verbindung von Industrialisierung, Urbanisierung und technischem Wandel, die gemeinsam die Mensch-Umwelt-Beziehungen dramatisch veränderten. Die Industrialisierung hatte mindestens zwei Konsequenzen. Zum einen erzeugte sie neue Probleme in Form von Verunreinigungen. Zum zweiten schuf sie aber auch neue Lösungsmöglichkeiten für Umweltprobleme, die sich etwa im Bereich der Städte ergaben. Jede technische Lösung zeitigte jedoch wiederum neue Probleme. Schott erläutert, dass Umweltprobleme in den großen Städten im Deutschland des späten 19. Jahrhunderts offen debattiert wurden, mehr oder weniger ausschließlich im Sinne der öffentlichen Gesundheit und der Ökonomie bzw. als Schutz des Eigentums. Die in Großbritannien ein wenig früher geführten Debatten über Hygiene und die dort vorgeschlagenen Lösungen hatten großen Einfluss auf die politischen Eliten und die technischen Experten in Deutschland. Die Wasserversorgung ist dafür eines der wichtigsten Beispiele. Die neu installierten Abwasserkanalisationen etwa waren höchst komplexe technische Infrastrukturen, die den Wasserkreislauf veränderten und die Lebensbedingungen in den Städten erheblich verbesserten; sie hatten jedoch enorme Schadstoffbelastungen in den großen Flüssen zur Folge. Das zweite von Schott angeführte Beispiel betrifft die Energieversorgung und die Versuche, die damit verbundene Luftverunreinigung mit Mitteln der Stadtplanung in den Griff zu bekommen. Dabei unterstreicht er die soziale Komponente aller Lösungsansätze, die in der Regel die Wohlhabenden begünstigte.

Die Umwelt schützen

Kein Überblick über Umweltgeschichte kann auf eine Erörterung von Natur- und Umweltschutz verzichten. In beiden Ländern, Deutschland wie Großbritannien, hat die Naturschutzbewegung eine lange Geschichte. *Ian Simmons* weist darauf hin, dass jede Gesellschaft, die natürliche Ressourcen nutzt, diese in der einen oder anderen Weise schützen muss. Sein Beitrag konzentriert sich auf die Epoche der Industrialisierung, wobei er insbesondere Maßnahmen in den Blick nimmt, die dazu dienten, Umweltverunreinigungen infolge industrieller Produktion zu begrenzen. Seit Mitte des 19. Jahrhunderts bemühte sich der britische Gesetzgeber, Industrieemissionen zu kontrollieren. Ab dem frühen 20. Jahrhundert wandte sich die Politik auch der Landnutzung als zweitwichtigstem Regelungsbereich zu, wobei die Planungsgesetze der vierziger Jahre die wohl wichtigsten Instrumente bereitstellten. Die Geschichte der Nationalparks seit dem Zweiten Weltkrieg und ihrer Verwaltung, der Nature Conservancy, zeigt die Bedeutung einer öffentlichen Fachverwaltung für die Umweltpolitik. In letzter Zeit hat schließlich die Europäische Kommission die administrative Schlüsselrolle im Umweltschutz übernommen. Die Ergebnisse von etwa anderthalb Jahrhunderten Umweltschutz sind nicht eindeutig. Viele Regelungen wurden durch Vollzugsdefizite und Interessenkollisionen in ihrer Wirkung beschnitten. Mit Blick auf die Schutzpraktiken schließt Simmons, dass das Prinzip selektiver Eindämmung der Belastungen nach wie vor dominiere, so dass die Vorstellung einer ökologischen Wende um 1970 sehr fraglich erscheint.

Kai Hünemörder betrachtet in seinem Beitrag die Geschichte des deutschen Naturschutzes. Er skizziert die wichtigsten Entwicklungen im Laufe des 20. Jahrhunderts, wobei er drei Hauptphasen unterscheidet, die er am Beispiel von Schleswig-Holstein untersucht. Der Artikel stellt die Interdependenz regionaler, nationaler und internationaler Weichenstellungen vor, verweist aber auch darauf, dass Natur letztlich nur vor Ort, auf lokaler Ebene, erfahrbar ist. Um 1900 kam in Deutschland die von den Mittelschichten dominierte Natur- und Heimatschutzbewegung auf und entfaltete ein Schutzkonzept. Die ersten Gesetze, Verwaltungsvorschriften und die Gründung der Staatlichen Stelle für Naturdenkmalschutz belegen die politische und kulturelle Wirksamkeit dieser zivilgesellschaftlichen Bewegung. Die Ära des Dritten Reiches erwies sich als sehr ambivalent für die Sache des Naturschutzes. Einerseits unterstützte das Naziregime den Naturschutz mit dem Erlass eines modernen Schutzgesetzes und der damit verbundenen Restrukturierung und Stärkung der Naturschutzverwaltung. Zudem förderte es die Berücksichtigung moderner Planungsinstrumente im Rahmen der Naturschutzkonzeptionen. Andererseits konterkarierten die Auswirkungen des Zweiten Weltkriegs und der Autarkiepolitik diese Neuerungen und verursachten erhebliche Schäden in der natürlichen Umwelt. Schließlich waren nicht wenige Naturschutzfunktionäre in Kriegsverbrechen verwickelt, die sie infolge der Umweltplanungen für die besetzten Gebiete begingen. Um 1970 veränderte die Idee der Ökologie das Gesicht des Naturschutzes, der aus seinem nationalen Bezugsrahmen heraustrat und zunehmend als internationale Angelegenheit betrachtet wurde. In Schleswig-Holstein ermöglichte das neue Umweltbewusstsein erstmals die Einrichtung großflächiger Schutzgebiete im Wattenmeer.

Prinz-Albert-Medaillen

Zum Abschluss präsentiert *David Allan* eine umfassende Darstellung der Geschichte der Prinz-Albert-Medaille, die erstmals 1863, zwei Jahre nach dem frühen Tod des Prinzen, an den Gründer des modernen Postwesens, Sir Robert Hill, verliehen wurde. Im Kontext der umwelthistorischen Tagung ist diese Geschichte deshalb so faszinierend, weil sie den Zusammenhang der Ziele der Royal Society of Arts (nämlich die Förderung von Wissenschaft, Kunst und Handel) mit Umweltfragen zeigt. Eine der grundlegenden Annahmen in vielen modernen Debatten geht davon aus, dass wirtschaftlicher und vor allem technischer Fortschritt der Wahrnehmung und Lösung von Umweltproblemen zuwiderlaufe. Tatsächlich zeigt die jüngste Auszeichnung von Tim Smit mit der Prinz-Albert-Medaille für seine Verdienste um die Ökologie im Rahmen des von ihm gestalteten „*Eden Projektes*", dass alle diese Elemente zusammenwirken können und sollten, um eine existenzfähige und nachhaltige Zukunft für die Menschheit und die Natur zu erzielen.

Fazit

Die Beiträge in diesem Buch bieten einen guten Überblick über Stand und Schwerpunkte der umwelthistorischen Debatten in Deutschland und Großbritannien. Es ist deutlich geworden, dass viele der umwelthistorischen Themen eng mit Fragestellungen aus der Allgemeinen Geschichte verbunden sind, was sich am Beispiel der Urbanisierung und der Industrialisierung ebenso zeigt, wie bei den rechtlichen und infrastrukturellen Bedingungen von Ressourcennutzung und Grundeigentum. Auch sollten die Potenziale einer Zusammenarbeit mit anderen Disziplinen deutlich geworden sein.

Die Tagung ermöglichte einen nützlichen, fruchtbaren Austausch und regte eine Reflexion über den Stand des Faches in den beiden Ländern an. Solche Gelegenheiten sind wichtig, weil sie es erlauben, von Zeit zu Zeit die Lücken und Schwächen genauso wie die Erfolge und die künftigen Forschungsaufgaben zu verdeutlichen. Vor allem aber hat die Konferenz gezeigt, dass Historiker einen wichtigen Beitrag zum Verständnis der komplexen Beziehungen zwischen Mensch und Umwelt in einer Langzeitperspektive leisten können – einen Beitrag, der auch für die künftige Gestaltung dieser Beziehungen wichtig ist. Schließlich gibt es keinen Zweifel über die immense politische und gesellschaftliche Bedeutung der Umweltfrage, die auch in den folgenden Jahrzehnten in der nationalen wie internationalen Politik eine zentrale Rolle spielen wird.

Christopher Smout

The Environmental Historiography of Britain

When did the British discover that the environment had a history? It was not in 1555, when the term "natural history" entered the English language, or in 1686 when John Ray published the first volume of his *Historia Generalis Plantarum*, or even in 1771, when Gilbert White sketched to Thomas Pennant his project for *"a natural history of my native parish"*, to become the *Natural History of Selbourne*. "History" as used in these titles meant "description" in the Aristotelian sense of scholarly knowledge. Natural history was by no means divorced from the Judaic view of nature expressed in the Book of Job and in the psalms, recovered for Christianity by St Thomas Aquinas and St Francis, and expressed in English sculpture and illustration between the end of the thirteenth and the start of the fifteenth centuries in the accurately observed foliage and animal carvings of Southwell Minster and the lively fishes swimming round the feet of the majestic Runcorn St Christopher or the charming native birds in the margins of the Sherborne missal. These are nature as delight and nature showing forth the glory of God[1].

But the natural history of later centuries was also essentially about description, a systematic account of a set of natural phenomena. It was divorced from the medieval fables of the bestiary, whereby creatures were set on earth to symbolise moral qualities, though this was to have a life of its own that flourished until Victorian times. Natural history was neither divorced from the modern project of discovering economic uses for nature, nor from the ancient one of showing forth the glory of the Creator, later rechristened natural religion[2]. But natural history had no time-line, so it was not history in our sense.

The British discovered environmental history during the Enlightenment. As good a date as any is 1788, when the Scottish geologist James Hutton, friend of Hume and Smith, published his *Theory of the Earth* and concluded a scientific lecture given before the Royal Society of Edinburgh with the scalp-tingling words, *"we find no vestige of a beginning – no prospect of an end"*. The notion that the world was only 6,000

[1] See *Nikolaus Pevsner*, The Leaves of Southwell, London 1945; *Janet Blackhouse*, Medieval Birds of the Sherborne Missal, London 2001.

[2] *Charles E. Raven*, English Naturalists from Neckam to Ray, Cambridge 1947; *Charles E. Raven*, John Ray Naturalist. His Life and Works, Cambridge 1950; *Keith Thomas*, Man and the Natural World: Changing Attitudes in England, Harmondsworth 1983.

years old, as reckoned from the Biblical generations, finally went out of the window then, though it had already been undermined – for instance by Newton's calculation that the molten sphere of iron that was the earth would take 50,000 years to cool. Even Hutton could not guess that the universe was 6000 million years old, not 6000 – that immensity would take until the twentieth century to discover[3].

Victorian geologists who wanted both to remain Christian and to respect the science of course knew through the study of strata and fossils that the environment was immensely ancient, and treated Genesis as a parable. One of the most influential was the very gifted Scottish amateur Hugh Miller, who argued that the seven biblical days of creation were in fact "prophetic days" of long duration for which the periods of geology provided evidence: six earlier days of primitive sea-life, of primitive forests, of birds and reptiles, of mammals and of man. The present day is the seventh or Sabbath day, when man must be redeemed, and with the rise of regenerate man, God's intention for creation will be complete[4]. Miller was responding to Robert Chambers' early and faith-eroding views of evolution expressed in *Vestiges of the Natural History of Creation* (1844), but with Wallace and Darwin, evolution became continuous and needed no invisible hand at all[5]. By the time of the publication of the *Origin of Species* in 1858 the battle lines were drawn between the creationists and the evolutionists. There is a sense in which environmental history became the central intellectual problem of the Victorian age[6].

It was not, however, an environmental history (in the sense that historians mean in the twenty-first century) that was actually happening around Victoria and Albert as they sat admiring the harvest fields on the Isle of Wight, or walked together in the ancient pine forest around Balmoral. It was the environmental history of the remote past, of the cosmos and the coal measures, of the fossil tree and the ammonite, the environmental history of physical geographers, which deals, or can deal, in thousands of millennia, and has no need for people to be on earth to have a history. "Environmental history" is used in that sense by British physical geographers to this day, and remains a productive and brilliant tradition. For example, in 1996 Edinburgh University hosted a symposium on the Environmental History of the Cairngorms in which

3 *D.R. Dean*, James Hutton and the History of Geology, Ithaca and New York 1992; *Donald B. McIntyre* and *Alan McKirdy*, James Hutton, the Founder of Modern Geology, National Museums of Scotland, Edinburgh 2001.

4 *M. Shortland*, Hugh Miller and the Controversies of Victorian Science, Oxford 1996; *Charles Waterston*, An Awakening Interest in Geology, Hugh Miller in Context, ed. Lester Borley, Cromarty 2002, 85-90. See also Celebrating the Life and Times of Hugh Miller, ed. *Lester Borley*, Cromarty 2003.

5 *James A. Secord*, Victorian Sensation. The Extraordinary Publication, Reception and Secret Authorship of Vestiges of the Natural History of Creation, Chicago 2000; *Peter Raby*, Alfred Russel Wallace. A Life, London 2001; *Peter J. Bowler*, Charles Darwin: The Man and His Influence, Cambridge 1990; *Janet Browne*, Charles Darwin, vol. 1: Voyaging, vol. 2: The Power of Place, London, New York 1995 and 2002.

6 *Peter J. Bowler*, Darwinism and Victorian Values, Threat or Opportunity: Victorian Values, A Joint Symposium of the Royal Society of Edinburgh and the British Academy December 1990, ed. T.C. Smout, (Proceedings of the British Academy 78) Oxford 1992, 129-147.

documentary history played no part at all, but which showed a dazzling array of scientific methodologies through which to explore the past. One paper used the analysis of the fossilised heads of non-biting midges deposited in high mountain lochs to indicate 3,000 years of climate change. Others used carbon and lead or zinc particles embedded in trees or deposited in water to study pollution over hundreds of years, and to show how changing levels of acidity had altered the diatom flora of fresh water, suddenly and abruptly in the twentieth century[7].

I dwell on this example (which is not a particularly unusual one of the way physical geographers' work) to emphasise the way in which scientists over the last century have added an extraordinary range of tools to understand the environmental past, many of which we historians are only dimly aware: in certain cases the scientists concerned have also combed documentary sources to add another and what is for us more recognisable and recognised perspective. Thus Peter Brimblecombe in his remarkable study of air pollution in London writes both as a scientist and an historian[8], and the work of the climate history school at Norwich pioneered by H.H. Lamb involves ice-cores, historical records of temperature and rainfall since the seventeenth century and the qualitative evidence of diaries and chronicles[9]. Mike Baillie in Belfast argues for the value of dendrochronology, the study of tree-rings, not only as an archaeological dating technique but also as a source for the history of climatic crises over at least two millennia[10]. Jean Grove's classic study of the Little Ice Age is a synthesis of much European as well as British climate history research in this tradition[11].

Of all the methodological contributions that environmental scientists have made, probably none has greater current significance for the study of environmental history than radiocarbon dating, used alongside palynology (the study of fossil pollen), two fields in which the main pioneer in Britain was Sir Harry Godwin of Cambridge University, who applied the techniques to develop the theory of ecological succession. His work is commemorated and furthered in the Godwin laboratory there. While early radio carbon dating was relatively crude, recent advances in that and allied dating techniques, using lead isotopes, have led to considerable refinement.

Palynology can be applied to previous interglacials where the muds and peats are old enough, but its greatest application has been to study the early and mid Holocene (the period since the last Ice Age), and through it we have gathered a fairly clear idea

7 Environmental History of the Cairngorms, Botanical Journal of Scotland 1 (1) (1996). See in particular *S.J. Brooks*, Three Thousand Years of Environmental History in a Cairngorms Lochan Revealed by Analysis of Non-biting Midges, 89-98; *N.J. Loader* and *V.R. Switsur*, Reconstructing Past Environmental Change Using Stable Isotopes in Tree-rings, 65-78; *R.W. Battarbee, V.J. Jones et al*, Palaeolimnological Evidence for the Atmospheric Contamination and Acidification of High Cairngorm Lochs, with Special Reference to Lochnagar, 79-88.

8 *Peter Brimblecombe*, The Big Smoke. A History of Air Pollution in London since Medieval Times, London 1987.

9 *H.H. Lamb*, Climate: Present, Past and Future, London 1972; *H.H. Lamb*, Climate, History and the Modern World, London 1982.

10 *M.G.L. Baillie*, A Slice Through Time: Dendochronology and Precision Dating, London 1995.

11 *Jean Grove*, The Little Ice Age, London 1988.

of the succession of forest communities in Britain and Europe that formed the environmental setting of the Mesolithic, the Neolithic, the Bronze Age and the Iron Age. It has helped archaeologists and early historians completely to revise their view of how wooded Scotland was, for example, at the time of the Roman invasion. In this context I am most cognisant of the work of Richard Tipping of Stirling, whose work on Scottish woodland history, and that of his students, has been of outstanding value to the historical and archaeological community[12].

A very recent example of the use of palynology, and of that of the comparable study of early beetle fauna from insect remains, has been to test Franciscus Vera's hypothesis that the original-natural wildwood of the post-glacial Mesolithic millennia had not been closed forest but open country in which the woods were groves surrounded by thorny edges, with grazing by large herbivores as the force behind cyclical vegetation dynamics and shifting mosaics of forest. The weight of evidence from the scientists assembled by Keith Kirby for English Nature has been to suggest that Vera's hypothesis, as stated, is overdrawn, and that Britain had a mixed landscape with some areas of vegetation cycles and others with more permanent closed forest, with herbivores not always the driving force[13].

Relatively little palynology has seriously overstepped the boundary of the last 1000 years except in one or two recent Ph.D theses, but it is fascinating for the documentary historian to see what these techniques might yield for the relatively recent past[14].

To palynology and all those other techniques I have mentioned there should be added the work of soil scientists, who seem to me to have it within their power one day, through the analysis of buried soils, to be able to throw much light not only on early manuring practices but also, for example, on questions of nutrient exhaustion in medieval agriculture[15]. Similarly genetic studies and the science of DNA have the potential to illuminate historical patterns of human and animal migration in ways at which we can barely guess.

I have emphasised the contributions of applied physical science and physical geography because I believe that all modern environmental historians can still learn so much from it. But modern environmental history has a variety of other roots, one of

12 *Richard Tipping*, The Form and Fate of Scottish Woodlands: Proceedings of the Society of Antiquaries of Scotland 124 (1994), 1-54; Living in the Past, Woods and People in Prehistory to 1000 BC: People and Woods in Scotland, ed. *T.C. Smout*, Edinburgh 2003, 14-39.

13 *F.W.M. Vera*, Grazing Ecology and Forest History, Oxford and New York 2000; *K.H. Hodder, J.M. Bullock, P.C. Buckland* and *K.J. Kirby*, Large Herbivores in the Wildwood and Modern Naturalistic Grazing Systems, (English Nature Report 648) 2005.

14 E.g. *Philip Sansum*, Historical Resource Use and Ecological Change in Semi-natural Woodland: Western Oakwoods in Argyll, Scotland, University of Stirling Ph.D Thesis 2004.

15 *E.I. Newman* and *P.D.A. Harvey*, Did Soil Fertility Decline in Medieval English Farms? Evidence from Cuxham, Oxfordshire, 1320-1340: Agricultural History Review 45 (1997), 119-136; *D.A. Davidson* and *I.A. Simpson*, Soils and Landscape History – case studies from the Northern Isles of Scotland, The History of Soils and Field Systems, eds. S. Foster and T.C. Smout, Aberdeen 1994, 66-74.

which is the interest that zoologists themselves felt in the way that their subjects of study had interacted with man in recent centuries. Victorian natural historians were increasingly interested in species history, especially where man had brought about or threatened extinction. The extinction of the Great Auk was traumatic to Victorian sensibilities and led the Cambridge zoologist Alfred Newton down the road towards legislative protection for seabirds, and it also led in 1885 to a remarkable species history of the Great Auk by Symington Grieve[16]. The Scottish naturalist J.A. Harvie-Brown wrote a series of species histories of taxa either at risk from local extinction or re-introduced after extinction, the Capercaillie, the Great Spotted Woodpecker and the Red Squirrel, the last of which led to a book that aroused the special attention of the Royal Family[17]. When natural history turned towards ecology, an interest grew in the interaction between biological groups or communities and man. It was evident early in the twentieth century by interest in the German concept of "Urwald" or wildwood, and how far it might apply in Britain[18]. After the First World War, James Ritchie, *Influence of Man on Animal Life in Scotland* (Cambridge 1920) was a milestone, ambitious and comprehensive, very modern in conception and execution but still underestimated. After the Second World War, the Collins New Naturalist series encompassed a number of volumes all by scientists, who, like Ritchie, were caught up in the emerging nature conservation movement and wished to write about the interactions between man and nature, inevitably with an historical perspective: R.S.R. Fitter, *London's Natural History* (London 1945), E.M. Nicholson, *Birds and Men: the Birdlife of British Towns, Villages, Gardens and Farmland* (London 1951) and H.L. Edlin, *Trees, Woods and Man* (London 1956), are good examples of this type of study[19]. It is a tradition that continues to provide distinguished work, as in Derek Yalden, *The History of British Mammals* (London 1999) and Michael Shrubb, *Birds, Scythes and Combines: A History of Birds and Agricultural Change* (Cambridge 2003), which provides a detailed and perceptive study of the interaction of agricultural history and birdlife since the eighteenth century. In the same lineage will be Roger Lovegrove's forthcoming history of the human treatment of birds in Britain, centred on exhaustive research on English church-wardens accounts of 'vermin' bounties since the reign of Tudors. Also of great interest is the floral history of Glasgow produced by Jim Dickson and his collaborators, showing how vibrant and dy-

[16] *Symington Grieve*, The Great Auk or Garefowl (*Alca impennis* Linn.): its History, Archaeology and Remains, Edinburgh 1885. More recent studies include *Errol Fuller*, The Great Auk, Southborough and Kent 1999; *W.R.P. Bourne*, The Story of the Great Auk *Pinguinus impennis*: Archives of Natural History 20 (2) (1993), 257-278.

[17] *J.A. Harvie-Brown*, The Capercaillie in Scotland, Edinburgh 1879; The History of the Squirrel in Great Britain: Proceedings of the Royal Physical Society, Edinburgh (1881); The Great Spotted Woodpecker (Picus Major, L.) in Scotland: Annals of Scottish Natural History 1 (1892), 4-17.

[18] *G.P. Gordon*, Primitive Woodland and Plantation Types in Scotland: Transactions of the Royal Scottish Arboricultural Society 24 (1911), 153-177.

[19] Though by a human geographer rather than a scientist another book in the same series is also environmental history: *L. Dudley Stamp*, Man and the Land, London 1964. It contains among other things a chapter on alien species.

namic the ecosystem of a great city can be, no whit less interesting for the role of so many despised 'alien species'[20].

The outstanding example of the tradition of scientist working as historian, however, is Oliver Rackham, probably the only environmental historian whose name is widely recognisable to the contemporary British public. Rackham is a Cambridge botanist by training who has acquired the skills of a medieval and early modern historian, and is a Fellow of the British Academy rather than of the Royal Society on account of his contribution to the humanities. His works on woodland history, of the Mediterranean as well as of England, have had international impact, and by defining and emphasising the concept of ancient woodland as a living category in the English countryside, he has had a profound practical impact on modern British conservation. Scarcely any forester in the British Isles can now think of his job without being aware of woods as historical artefacts and of the work of Oliver Rackham in sensitising the public to their history[21]. He began his notable studies with detailed work on a local wood in Cambridgeshire, *Hayley Wood: its History and Ecology* (Cambridge 1975) but his most influential works on British woodland history have been *Trees and Woodland in the British Landscape: the Complete History of Britain's Trees, Woods and Hedgerows* (London 1976), and *Ancient Woodland, its History, Vegetation and Uses in England* (London 1980: new edition Colvend, Kirkudbrightshire 2003). It was, however, his *History of the Countryside* (London 1986), a popular and general but never condescending work, that first excited me about the intellectual possibilities of environmental history.

It is interesting that Rackham has been a consistent critic of the sort of environmental history that lays more emphasis on the history of people using and thinking about nature than about the history of natural things used by people: as he accurately put it concerning a volume that I edited, a book entitled *People and Woods* is more likely to be mainly about people than mainly about woods. It is fair comment, but since the 1970s various sorts of British historians, and no longer just scientists, have interested themselves in environmental history, and they do also tend to be seriously interested in people. The rise of a new political and intellectual 'environmentalism' inspired some to explore the origin of changes in attitudes towards nature and the rise of conservation politics. Keith Thomas of Oxford, polymath among English historians, produced his outstanding and influential volume, *Man and the Natural World: Changing Attitudes in England* (Harmondsworth 1983), tracing how English ideas about plants and animals developed between 1500 and 1800. To a degree it had been paralleled and anticipated in literary studies a decade earlier by Raymond Williams's equally brilliant *The Country and the City* (London 1973). John Sheail, trained as a

20 *J.H. Dickson, P. Macpherson* and *K. Watson* (with contributions from others), The Changing Flora of Glasgow. Urban and Rural Plants through the Centuries, Edinburgh 2000.

21 *Judith Tsouvalis*, A Critical Geography of Britain's State Forests, Oxford 2000; *J. Tsouvalis-Gerber*, Making the invisible visible: ancient woodlands, British Forest Policy and the Social Construction of Reality, European Woods and Forests: Studies in Cultural History, ed. Charles Watkins, Wallingford 1998.

Tudor historian but working in nature conservation agencies, produced a range of seminal volumes about the history of ecology, conservation and environmental planning in Great Britain[22]. D.E. Allen wrote a memorable history of British naturalists in 1976, emphasising in particular the distinctive national amateur tradition[23].

Others, notably economic historians and human geographers, began to examine resource history in the light of environmental concerns and interests. In 1994 Brian Clapp wrote a sketchy but pioneering overall view of the environmental history of Britain since the Industrial Revolution from a rather narrow economic historian's point of view[24]. Much more ambitious and successful was the work of Ian Simmons, who as a geographer in 2001 capped many years of scholarship exploring environmental history (from the archaeological perspective among others), with a magisterial overview of British environmental history as a matter of resource use[25]. Angus Winchester explored the environmental history of the northern English and southern Scottish uplands very much with the eye of an economic and agrarian historian interested in allocation and community division of resources[26]. At the University of Oxford, the annual Linacre lectures on the environment have since 1997 resulted in seven volumes which surprisingly seldom deal with environmental history and still less often with the British experience of it[27]. On the other hand, The Institute for Environmental History at the University of St Andrews produced between 1993 and 2002 a series of books based on conference proceedings, ranging from soil history and species history to landscape history, mainly, though not exclusively, with a Scottish emphasis[28]. In my own Ford Lectures in Oxford in 1999 I tried to suggest a tension between use of nature and a delight in dealing with nature, in the specific context

[22] *John Sheail*, Nature in Trust, the History of Nature Conservation in Britain, London and Glasgow 1976; Rural Conservation in Interwar Britain, Oxford 1981; Pesticides and Nature Conservation: The British Experience 1950-1975, Oxford 1985; Seventy-five Years in Ecology, The British Ecological Society, Oxford 1987; Nature Conservation in Britain – The Formative Years, London 1998; An Environmental History of Twentieth-Century Britain, Basingstoke 2002.

[23] *David Elliston Allen*, The Naturalist in Britain, a Social History, London 1976.

[24] *Brian W. Clapp*, An Environmental History of Britain since the Industrial Revolution, London 1994.

[25] *I.G. Simmons*, An Environmental History of Great Britain from 10,000 Years Ago to the Present, Edinburgh 2001. See also The Environment in British Prehistory, eds. *Ian Simmons* and *Michael Tooley*, Cornell 1981.

[26] *Angus J.L. Winchester*, The Harvest of the Hills, Rural Life in Northern England and the Scottish Borders, 1400-1700. See also *I.G. Simmons*, The Moorlands of England and Wales, an Environmental History 8000BC – AD2000, Edinburgh 2003.

[27] The most relevant are Culture, Landscape and the Environment, eds. *Kate Flint* and *Howard Morphy*, Oxford 2000 and Environments and Historical Change, ed. *Paul Slack*, Oxford 1999.

[28] Scotland Since Prehistory: Natural Change and Human Impact, ed. *T.C. Smout*, Aberdeen 1993; The History of Soils and Field Systems, eds. *S. Foster* and *T.C. Smout*, Aberdeen 1994; Fragile Environments, the Use and Management of Tentsmuir NNR, Fife, ed. *Graeme Whittington*, Edinburgh 1997; Scottish Woodland History, ed. *T.C. Smout*, Edinburgh 1998; Species History in Scotland, Introductions and Extinctions since the Ice Age, ed. *Robert A. Lambert*, Edinburgh 1998; Rothiemurchus: Nature and People on a Highland Estate, 1500-2000, eds. *T.C. Smout* and *R.A. Lambert*, Dalkeith 1999; Understanding the Historical Landscape in its Environmental Setting, ed. *T.C. Smout*, Dalkeith 2002.

of the environmental history of Scotland and Northern England since 1600[29]. The same theme was taken up by Robert Lambert in a study of conflict over land use in the Cairngorms from 1880-1980[30]. Others, notably Charles Watkins, Richard Muir, William Linnard, Ian Rotherham and members of the Scottish Woodland History Discussion Group, built on the work of Rackham to examine new areas of woodland history both in the geographical and intellectual sense[31]. Stephen Mosley followed Brimblecombe's lead with a study of smoke pollution in Victorian and Edwardian Manchester, but with more focus on social and political urban history[32].

So far, perhaps to your surprise, I have spoken without reference to the influence of the great efflorescence of environmental history that occurred in America from the 1970s onwards, which has produced such notable authors as Donald Worster, William Cronan, J.R. McNeill, Roderick Nash, Caroline Merchant, Stephen Pyne, Alfred Crosby and Jahred Diamond[33]. The list is much longer than I can indicate here, and in many respects the American school dominates the subject, so that today what is usually meant by 'environmental history' internationally is the humanities-based and often ideologically-committed subject associated with the Americans. Of course it has had enormous influence in Britain well as elsewhere, but most notably it has been among historians working in British universities but studying other continents: Richard Grove, for instance, on tropical islands and on India, William Beinart and John MacKenzie on Africa, Mark Elvin on China, Peter Coates on Alaska, to name

29 *T.C. Smout*, Nature Contested, Environmental History in Scotland and Northern England since 1600, Edinburgh 2000.

30 *Robert A. Lambert*, Contested Mountains, Nature Development and the Environment in the Cairngorms Region of Scotland 1880-1980, Cambridge 2001.

31 The Ecological History of European Forests, eds. *Keith J. Kirby* and *Charles Watkins*, Wallingford 1998; European Woods and Forests, Studies in Cultural History, ed. *Charles Watkins*, Wallingford 1998; *Richard Muir*, Ancient Trees, Living Landscapes, Stroud 2005; *William Linnard*, Welsh Woods and Forests, a History, Llandysul and Ceredigian 2000; Ancient Woods, their Archaeology and Ecology, eds. *P. Beswick* and I. *Rotherham*, Sheffield 1993; People and Woods in Scotland, a History, ed. *T.C. Smout*, Edinburgh 2003; *T.C. Smout*, *Alan R. MacDonald* and *Fiona Watson*, A History of the Native Woodlands of Scotland, 1500-1920, Edinburgh 2005. The Scottish Woodland History Discussion Group Notes have been published annually, based on conference papers, since 1996 (ISSN 1470-0271). In Scotland there is an independent older tradition of woodland history written by foresters going back to *H.M. Steven* and *A. Carlisle*, The Native Pinewoods of Scotland, Edinburgh 1959 and *Mark Louden Anderson*, A History of Scottish Forestry (2 vols.), London 1967.

32 *Stephen Mosley*, The Chimney of the World: A History of Smoke Pollution in Victorian and Edwardian Manchester, Cambridge 2001.

33 The following is a mere sample of their work: *Donald Worster*, Nature's Economy, a History of Ecological Ideas, Cambridge 1977; *William Cronan*, Nature's Metropolis: Chicago and the Great West, New York 1991; *J.R. McNeill*, Something New Under the Sun: an Environmental History of the Twentieth-Century World, New York 2000; *Roderick Nash*, Wilderness and the American Mind, New Haven 1982 (third edn.); *Carolyn Merchant*, Ecological Revolutions, Nature, Gender and Science in New England, Chapel Hill 1989; *Stephen J. Pyne*, Fire in America, a Cultural History of Wildland and Rural Fire, Princeton 1982; *Alfred W. Crosby*, Ecological Imperialism, the Biological Expansion of Europe, 900-1900, Cambridge 1986; *Jared Diamond*, Collapse, How Societies Choose to Fail or Survive, London 2005.

only some of the most distinguished[34]. This is mainly because, as defined by the American school, environmental history has essentially been an extension of the history of imperial expansion, of colonial impacts with European crops, animals and machinery on fragile environments, whether these be the mid-western plains, Caribbean islands, the tropics or the South African veldt. The questions they were asking seemed much less appropriate to a country (indeed, to a continent) where invasion and overthrow are not usually seen in the same imperial expansionist terms, and where the environment both appears superficially to be much more robust (at least in terms of soil and water) and where, perhaps more pertinently, crops and animals evolved over centuries to grow, more or less sustainably, within it.

The history of the environment seemed less to be environmental history in the American sense of a history of misunderstanding and violent misuse, as a history of relatively benign and gradual landscape change and of an agriculture that by certain definitions at least was environmentally sustainable until after the Second World War. Landscape history and agrarian history are subjects in which British geographers and historians have been distinguished since the days of H.C. Darby and W.G. Hoskins[35], a tradition which is continued today by cultural geographers like David Lowenthal and historical geographers like Robert Dodgshon[36]. As might be expected, geographers as a profession have had a particular sensitivity to environmental history, exemplified by two volumes published by the Institute of British Geographers in 1976 and 1995[37]. Journals on landscape history[38] also often contain more British environmental history than the two periodicals explicitly devoted to environmental history, the American-based *Journal of Environmental History* and the British-based *Environment and History.* Neither could be said to discriminate against British material, yet they fail to attract it as often as could be expected. The other route is the kind of economic history exemplified in the multi-volume *Agrarian History of England and*

34 *Richard H. Grove*, Green Imperialism: Colonial Expansion, Tropical Island Edens and the Origins of Environmentalism 1600-1960, Cambridge 1995; *William Beinart*, The Rise of Conservation in South Africa: Settlers, Livestock and the Environment, 1770-1850, Oxford 2003; *John M. MacKenzie*, The Empire of Nature: Hunting, Conservation and British Imperialism, Manchester 1988; *Mark Elvin*, the Retreat of the Elephants, an Environmental History of China, London 2004; *Peter A. Coates*, The Trans-Alaska Pipeline Controversy: Technology, Conservation and the Frontier, Fairbanks 1991; *William J. Beinart* and *Peter A. Coates*, Environment and History: the Taming of Nature in the USA and South Africa, London 1995. This is only a sample of the works of these authors.

35 *H.C. Darby*, The Domesday Geography of England (7 vols.), Cambridge 1952-1977; *W.G. Hoskins*, The Making of the English Landscape, London 1955.

36 *David Lowenthal*, The Past is a Foreign Country, Cambridge 1985; Landscape Meaning and Values, eds. *E.C. Penning-Rowsell* and *David Lowenthal*, London 1986; *Robert A. Dodgshon*, Land and Society in Early Scotland, Oxford 1981; Society in Time and Space: a Geographical Perspective, Cambridge 1998; Budgeting for Survival, Nutrient Flow and Traditional Highland Farming: The History of Soils and Field Systems, eds. *S. Foster* and *T.C. Smout*, Aberdeen 1995, 83-93.

37 Man's Impact on Past Environments, eds. *L.F. Curtis* and *I.G. Simmons*, Transactions of the Institute of British Geographers, n.s. 1 (1976), no. 3; Ecological Relations in Historical Times, eds. *Robin A. Butlin* and *Neil Roberts*, (Special Publications 32) Oxford 1995.

38 The journals are Landscape History and Landscapes. The Agricultural History Review also often contains articles of relevance to British environmental history.

Wales from pre-history to 1939, under the general editorship of Joan Thirsk, providing an achievement of truly monumental scholarship[39]. Joan Thirsk's own work – for example her brilliant history of alternative agriculture in England since the Black Death – in several respects is a striking example of environmental history by another name[40].

However, agricultural and landscape history are not themselves exactly synonymous with environmental history, as in them people and not the environment itself are centre stage. The sole example I can think of where an historian explicitly asks how fragile was the British environment in the face of industrial and agrarian change is James Winter's fine study of the impact on the Victorian countryside of the huge contemporary changes in economy and society[41]. Significantly, Winter's home base is the University of British Columbia in Vancouver. Yet Winter's book seems to be virtually unknown to mainstream scholars of nineteenth-century Britain. This points to a further weakness that Sverker Sörlin and Paul Warde of Cambridge have recently urged us to consider, namely that most environmental historians in Britain and indeed especially of Britain, seem to live in a ghetto where they do not interact much with other historians, and they are largely ignored by them in turn[42]. Another interesting question is the relationship between urban history and environmental history in Britain: at my own University of St Andrews a project on the history of urban waste is labelled and funded as environmental history but it could just as easily have been funded as urban history or social history. Perhaps this ambiguity is a strength, after all, and could help to overcome the ghetto mentality by a kind of permeability with other forms of history.

What of the future? I can sum up my impressions in three sentences, Firstly, we should celebrate that such a lot is going on especially among young scholars, Ph.D students and lecturers, in however disparate a fashion, because out of all this intellectual ferment new ideas and concepts will spring more readily than if we were all old, over-focussed and self-constrained. Secondly, we should not worry that the environmental history of Britain (or of Europe) is not necessarily about the same things as that of America, Australia or India, but, as Winter has shown, we could certainly learn to ask more questions about environmental sustainability and fragility. Lastly, we should build much more on interdisciplinary co-operation with our scientific colleagues, because it is here especially that British environmental history has actually been at its most pioneering and valuable, not just since the 1970s, but from Victorian times.

39 Agrarian History of England and Wales, gen. ed. *Joan Thirsk* (8 vols.), Cambridge 1967-2000.

40 *Joan Thirsk*, Alternative Agriculture. A History from the Black Death to the Present Day, Oxford 1997.

41 *James Winter*, Secure from Rash Assault. Sustaining the Victorian Environment, Berkeley 1999.

42 *Sverker Sörlin* and *Paul Warde*, The Problem of the Problem of Environmental History – a Re-reading of the Field and Its Purpose, paper delivered at Thinking through the Environment, VIII Turku Methodolical Conference/VI Nordic Environmental History Conference, Turku, Finland, 15 September 2005.

Franz-Josef Brüggemeier

Umweltgeschichte in Deutschland.

Ein Überblick

Die Umweltgeschichte hat in den letzten Jahren in Deutschland und international einen deutlichen Aufschwung erlebt, nicht nur weil die Zahl der einschlägigen Veröffentlichungen zugenommen hat. Deren Anstieg ist schon seit einiger Zeit zu beobachten, doch die Untersuchungen haben lange über den überschaubaren Kreis der Umwelthistoriker/innen hinaus wenig Beachtung gefunden. Diese Situation hat sich geändert. Die Umweltgeschichte findet mittlerweile mehr und mehr Einzug in etablierte Zeitschriften und Standardwerke und erlebt zudem einen Prozess der Historisierung, bei dem nicht nur unser Umgang mit der Umwelt, sondern auch die dazu geführten Debatten und die vorliegenden Veröffentlichungen untersucht und (selbst) kritisch bewertet werden.

Es gibt also Erfreuliches zu berichten, doch zugleich ist festzuhalten, daß nicht alle Erwartungen, die mit der Umweltgeschichte verbunden waren und noch verbunden sind, erfüllt wurden – wie dieser Überblick zeigen soll. Er wird dazu drei Bereiche behandeln und zuerst die Anfänge und Entwicklungen der Umweltgeschichte in Deutschland sowie die Impulse aus den USA skizzieren; anschließend fragt er, ob es sich hierbei um eine eigene historische Disziplin mit einer spezifischen Methode handelt, welche Rolle die seit einiger Zeit viel diskutierte Interdisziplinarität spielt und welche institutionelle Verankerung besteht; und schließlich gibt der letzte Teil einen Überblick über Themen und Fragestellungen, die derzeit vorherrschen.

1. Anfänge und Entwicklungen

Das Verhältnis von Mensch und Natur ist ein altes Thema der Geschichtswissenschaften, bei dem traditionell vor allem danach gefragt wurde, welche Einflüsse die natürliche Umwelt auf Lebensumstände, Entscheidungen und Verhaltensweisen von Menschen ausübt. Im Laufe des 19. Jahrhunderts veränderte sich die Sichtweise, als zunehmend deutlich wurde, welche Auswirkungen die Industrialisierung und der rasche Bevölkerungsanstieg ihrerseits auf Natur und Umwelt besaßen. Hierüber entstanden heftige Debatten, die notgedrungen eine historische Dimension besaßen, da es um Veränderungen durch menschliches Eingreifen ging. Explizite historische Untersuchungen hingegen entstanden noch nicht. Dazu kam es in Deutschland erst in den 1980er Jahren, als die Belastungen offensichtlich deutlich zugenommen hatten, Grenzen des Wachstums beschrieben wurden, Sorgen um ein Sterben der Wälder vorherrschten und zahlreiche andere Indikatoren auf eine dramatische Umweltkrise hinwiesen. Daß die Arten starben, die Meere verseucht waren und Klimakatastrophen drohten, gehörte *„zum Grundbestand des zeitgenössischen Bewußtseins“*[1]. Das Gleichgewicht der Erdatmosphäre galt (und gilt) als gefährdet, die befürchtete Aufwärmung werde *„in ihren Auswirkungen alle bisher bekannten Umweltkatastrophen übertreffen“*[2]. Die Welt müsse eine ökologische Revolution beginnen, oder sie riskiere *„einen ökologischen Kollaps und einen wirtschaftlichen Niedergang“*[3]. Die Debatte über die Bedeutung ökologischer Aspekte – so hieß es – war zu einer Diskussion *„über Tod und Leben“*[4] geworden.

Angesichts dieser Dramatik kam die Frage auf, ob ein Blick nach hinten überhaupt sinnvoll war oder angesichts der drängenden Probleme nicht vielmehr eine Zeitverschwendung bedeutete? So äußerste sich zumindest ein Mitarbeiter des Freiburger Öko-Instituts, der angesichts der *„tickenden ökologischen Zeitbomben“*[5] in historischen Lernprozessen eher eine Last sah. Derartige Meinungen waren verbreitet, und es ist deshalb nicht überraschend, daß die damalige Umweltdebatte vielfach ahistorisch verlief oder zumindest historische Arbeiten unter einen hohen Legitimationsdruck stellte. Die Krise erschien so groß, daß wir *„mit kleineren Korrekturen“*[6] offensichtlich nicht mehr auskamen, und entsprechend ehrgeizig fielen Fragestellungen und Konzepte aus. Im Kern ging es darum, die historischen Ursachen der Umweltkrise zu bestimmen und nach vergangenen Fehlentwicklungen und Sackgassen zu suchen. Dazu sollte sich die historische Forschung explizit an den Erkenntnissen der Ökologie und dem Konzept des Ökosystems orientieren[7]. Eine interdisziplinäre Zusammenarbeit galt als unerläßlich, ebenso die Überwindung des Anthropozentrismus, der den Menschen zu lange in den Mittelpunkt gestellt habe.

In der Frühzeit der modernen Umweltgeschichte dominierten Ansätze aus den USA. Daher ist es auch für einen Beitrag über Deutschland unerlässlich, die wichtigsten Impulse und Interpretationsansätze aus Nordamerika zu nennen. Als eine wesentliche Ursache der heutigen Bedrohung galt diesseits wie jenseits des Atlantik anfangs etwa der Auftrag der Bibel, die Menschen sollten sich die Erde untertan ma-

1 *Ulrich Beck*, Politik in der Risikogesellschaft. Essays und Analysen, Frankfurt/M. 1991, 9.

2 Das Ende des blauen Planeten? Der Klimakollaps: Gefahren und Auswege, hg. v. *Paul J. Crutzen* und *Michael Müller*, München ²1990, 2.

3 *World Watch Institute*, Zur Lage der Welt 1992. Daten für das Überleben unseres Planeten, Frankfurt/M. 1992, 7.

4 Mit den Bäumen sterben die Menschen. Zur Kulturgeschichte der Ökologie, hg. v. *Jost Hermand*, Köln/Weimar/Wien 1993, VI.

5 Zitat nach *F. Brenneke* und *P. Kern*, Neue Wege zur Umweltgeschichte, Manuskript 1994 (Forschungsinstitut für Arbeiterbildung, Recklinghausen), 1.

6 World Watch Institute, Zur Lage der Welt 1992. Daten für das Überleben unseres Planeten, Frankfurt/M. 1992, 10.

7 *Paul Leidinger*, Umwelterziehung im Geschichtsunterricht, in: Handbuch Praxis der Umwelt- und Friedenserziehung, hg. v. Jörg Calließ und Reinhold E. Lob, Bd. 2, Düsseldorf 1987, 281-92.

chen (dominium-terrae-Gebot). Im Gegensatz zum heidnisch-animistischen Weltbild, das jeden Busch oder Baum als Sitz eines Gottes betrachtet und dementsprechend geschont habe, sei hier die Wurzel der modernen Natur-Zerstörung zu suchen[8]. Das klingt beeindruckend, ist jedoch zu Recht kritisiert worden[9]. So hat gerade die römisch-katholische Kirche versucht, die Ausbreitung der modernen Naturwissenschaften zu verhindern. Auch sind die Personen, Gruppen und Gesellschaften, die sich auf die Bibel berufen, so zahlreich und verhalten sich so heterogen, daß es unmöglich ist, in diesem Auftrag der Bibel eine Gemeinsamkeit zu konstruieren. Zudem ist umstritten, wie diese Bibelstellen überhaupt zu verstehen sind. Bei richtiger Exegese – so andere Autoren – könne von einem Auftrag zur Ausbeutung der Natur keine Rede sein; gefordert werde vielmehr ein schonender Umgang mit der Natur.

Ebenso problematisch sind Arbeiten, denen zufolge die entscheidende Weichenstellung mit dem Übergang zur modernen Naturwissenschaft erfolgte. In den Worten von Carolyn Merchant, der wohl bekanntesten amerikanischen Umwelthistorikerin: *„Die Welt, in der wir leben, haben uns Isaac Newton und Gottfried Wilhelm von Leibniz vermacht“*[10]. Mit Newton und Leibniz habe sich eine mechanistische, männliche Weltauffassung durchgesetzt, die Natur sei als nützliches, totes Objekt begriffen worden. Ältere, ganzheitliche und organische Auffassungen der Natur seien verdrängt und Produktion sowie Reproduktion zunehmend voneinander getrennt worden, bis schließlich die „naturverhafteten“ Frauen auf die Reproduktion beschränkt worden seien. Die Sphäre der Produktion habe sich zu einem Betätigungsfeld für die Männer entwickelt und zunehmend aggressive Züge herausgebildet, bis hin zur heutigen Zerstörung der Natur. Die Frauen- und die Umweltbewegung sieht sie deshalb als enge Verbündete, die *„mit aller Schärfe gegen die Kosten* [*protestieren*], *die uns aus dem Konkurrenzprinzip, dem Aggressions- und Dominanzstreben des marktwirtschaftlichen Operierens in Natur und Gesellschaft erwachsen“*[11].

Diese Argumentation sieht die Geschichte der letzten 300 Jahre als Weg in eine Sackgasse, auf den die westlichen Kulturen sich unter dem Einfluß einiger (männlicher) Denker begeben haben. Das mag bedenkenswert sein, und derartige Überlegungen sind unter den Vertreter(inne)n radikal-ökologischer und feministisch-ökologischer Ansätze weit verbreitet, doch es fällt schwer, darin mehr als eine provokante These zu sehen[12]. So ist schon strittig, ob die Vordenker der Moderne – vor al-

[8] *Lynn White*, The Historical Roots of Our Ecological Crisis, Science 155 (1967), 1203-07; *Carl Amery*, Das Ende der Vorsehung. Die gnadenlosen Folgen des Christentums, Reinbek 1972.

[9] Vgl. *Udo Krolzik*, Umweltkrise – Folge des Christentums? Stuttgart [2]1980; *Gerhard Liedke*, Im Bauch des Fisches. Ökologische Theologie, Stuttgart 1979.

[10] *Carolyn Merchant*, Der Tod der Natur. Ökologie, Frauen und neuzeitliche Naturwissenschaft, München 1987, 263.

[11] Ebd., 12; ferner *dies.*, Gender and Environmental History, in: Journal of American History 76 (1990), 1117-1121.

[12] Vgl. *Barbara Holland-Cunz*, Utopien der Neuen Frauenbewegung: Gesellschaftsentwürfe im Kontext feministischer Theorie und Praxis, Meitingen 1988; Klasse Geschlecht: Feministische Gesellschaftsanalyse und Wissenschaftskritik, hg. v. *Ursula Beer*, Bielefeld 1987; *Evelyn Fox-Keller*, The Gender/Science System: Is Sex to Gender as Nature is to Science?, in: Hypatia 2 (1987), 37-49; *Ilona Ostner* und *Klaus Lichtblau*, Feministische Vernunftkritik. Ansätze und Traditionen, Frankfurt/M. 1992.

lem Bacon wird hier genannt – tatsächlich ein derart mechanistisches Weltbild besaßen und welche Bedeutung entsprechende Äußerungen überhaupt hatten. Thomas S. Kuhn weist darauf hin, daß die Männer, die im 17. Jahrhundert die wissenschaftliche Theorie umwälzten, zwar wie Bacon sprachen, doch es müsse noch nachgewiesen werden, „*daß die Ideologie, die eine Vielzahl von ihnen vertrat, eine größere substantielle oder methodologische Bedeutung für ihre zentralen Beiträge zur Wissenschaft besaß*“[13]. Hinzu kommen andere Probleme. So ist schwer nachzuvollziehen, wie nach den methodologischen Debatten der letzten Jahrzehnte einzelnen Personen eine derart epochale Bedeutung zugewiesen werden kann. Zu fragen wäre vielmehr, welche unterschiedlichen Faktoren und Konstellationen dazu beigetragen haben, daß sich das moderne Weltbild durchsetzen konnte, worin dessen Attraktion bestand, ob es Alternativen gab oder wie diese aussahen.

Als Antwort auf derartige Fragen wird oftmals auf vorindustrielle Gesellschaften verwiesen, die in größerer Harmonie mit der Natur gelebt hätten; oder auf die besondere Affinität von Weiblichkeit und Natur, doch diese Antworten sind inhaltlich und methodisch mehr als fragwürdig[14]. Inhaltlich sind sie eine Neuauflage der These vom goldenen Zeitalter, das als Folge eines Sündenfalls (mechanisches Weltbild, Christentum etc.) unterging, und methodisch schreiben sie Natur und Weiblichkeit a priori spezifische (positive) Qualitäten zu, die als Basis einer besseren Welt dienen sollen. Das ist schon deshalb überraschend, weil die Frauenbewegung der letzten Jahrzehnte einen großen Teil ihrer Energien gerade darauf gerichtet hat, die in diesem Ansatz enthaltenen Konzepte von Weiblichkeit (Mutterschaft, Passivität, Wärme etc.) als soziale Konstrukte zu entschlüsseln[15]. Hinzu kommt, daß die gepriesenen vorindustriellen Gesellschaften bei näherer Betrachtung viel von ihrer Attraktivität verloren haben. Es ist offensichtlich, daß Jäger- und Sammlergesellschaften die sie umgebende Umwelt und Natur nur wenig veränderten. Doch ihr Verhalten erklärt sich in erster Linie aus einer sehr geringen Bevölkerungsdichte, weniger aus einer besonderen Achtung vor der Natur. Der Umgang mit dieser konnte vielmehr äußerst verschwenderisch sein, wenn es etwa zu Brandrodungen kam oder Tierherden über Klippen getrieben wurden[16].

[13] *Thomas Kuhn*, The Essential Tension, Chicago 1977, 116 f.

[14] Als Beispiele: *Carolyn Merchant*, Ecological Revolutions. Nature, Gender, and Science in New England, Chapel Hill/London 1989; *Vito Fumagalli*, Mensch und Umwelt im Mittelalter, Berlin 1992.

[15] Hinzu kommt, dass implizit ein idealer Naturzustand bzw. eine essentielle Natur postuliert und darüber hinaus angenommen wird, diese sei weiblicher Erkenntnis in besonderer Weise zugänglich; vgl. *Evelleen Richards* und *John Schuster*, The Feminine Method as Myth and Accounting Ressource, in: Social Studies of Science 19 (1989), 697-720, und *dies.*, So What's not a Social Category?, *ebd., 725-729; Evelyn Fox-Keller*, Just what is so Difficult About the Concept of Gender as a Social Category, *ebd.*, 721-724; *Mary Mellor*, Breaking the Boundaries: Towards a Feminist, Green Socialism, London 1992, und *dies.*, Eco-Feminism and Eco-Socialism: Dilemmas of Essentialism and Materialism, Capitalism, in: Nature and Socialism 10 (1992), 43-62.

[16] Vgl. *Ian G. Simmons*, Environmental History. A Concise Introduction, Oxford 1993, Kap. 1 und 2.

Die Bemühungen um eine umfassende *„anthropologisch-ökologische Perspektive“*[17] und ein ‚ökologisches Paradigma‘ haben anfangs eine gewisse Faszination ausgeübt, die Umweltgeschichte jedoch in eine Sackgasse geführt. Ihre Annahme, es gebe stabile, ökologisch ideale Gleichgewichte, die erst menschliches Eingreifen gefährde oder gar vernichte, entsprach schon damals nicht dem Stand der naturwissenschaftlichen Forschung. Auch ohne menschliche Eingriffe befinden sich Ökosysteme in labilen Gleichgewichten, die durch innere oder äußere Kräfte gestört werden können, sich jeweils neu einpendeln. Sie kennen keine idealen Zustände und können entsprechend keine ‚objektiven‘ Maßstäbe bereitstellen, um menschliche Eingriffe in die Natur zu bewerten.

Das gilt umso mehr, als die natürliche Umwelt in Europa an nahezu keiner Stelle mehr ‚unberührt‘, d.h. nicht vom Menschen gestaltet ist. Es wird zudem immer schwieriger, zwischen natürlicher und künstlicher Natur zu unterscheiden. Wie verhält es sich mit einem eingehegten oder gar ‚renaturalisierten‘ Bach? Ist er noch natürlich? Oder ein von Menschen errichteter und von ihnen bepflanzter Hügel? Was ist mit genetisch manipulierten Pflanzen? Es scheint klar zu sein, daß eine industrielle und städtische Umwelt wie im Ruhrgebiet nicht natürlich ist. Doch wo läßt sich die Grenze ziehen? Und: Ist eine industriell geprägte Umwelt minderwertiger als eine wie immer zu bestimmende natürliche Umwelt?[18] Eine ‚eigentliche‘ Natur und objektive Maßstäbe zur Beurteilung dieser Frage gibt es nicht. Die Natur bzw. Umwelt, die wir uns wünschen oder vorstellen, ist das Ergebnis individueller oder kollektiver Werturteile.

Bei der Auseinandersetzung um diese Begriffe läßt sich ein Anthropozentrismus nicht vermeiden, ohne deshalb allerdings Menschen zum Maß aller Dinge machen zu müssen. Schon erkenntnistheoretisch ist kein anderer Zugang möglich. Menschen können die Umwelt nur aus ihrer Perspektive wahrnehmen, sie können sich nicht in die Position von Tieren und Pflanzen oder gar anorganischer Substanzen versetzen. Selbst wenn dies möglich wäre: Welche Position sollen sie einnehmen? Die von Füchsen? Das ginge vermutlich zu Lasten der Gänse. Oder die von Gänsen? Derartige Positionen werden nicht explizit vertreten[19], und fraglos spitzen diese Formulierungen das Problem zu[20]. Doch sie machen deutlich, daß ein allgemeines, für unser Handeln verbindliches Bezugssystem a priori nicht existiert, daß dieses vielmehr von Menschen geschaffen werden muß. Weder die Natur noch die Ökologie stellen allgemein-

[17] *Rolf P. Sieferle*, Perspektiven einer historischen Umweltforschung, in: Fortschritte der Naturzerstörung, hg. v. dems., Frankfurt/M. 1988, 351.

[18] Diese Frage hat unmittelbare politische Relevanz in der Debatte darüber, welchen Weg die Länder der Dritten Welt einschlagen sollen. Sollen sie ihre Natur ebenso umgestalten, wie es in der nördlichen Hemisphäre geschehen ist? Dagegen protestieren pikanterweise vor allem die industrialisierten Länder, die Nutznießer der bisherigen Umgestaltung.

[19] Implizit klingen derartige Vorstellungen bei Vertretern der „deep ecology“ an; vgl. *David Oates*, Earth Rising. Ecological Belief in the Age of Science, Corvallis 1989.

[20] Die Versuche, eine neue ökologische Ethik und darauf gründende politische, ökonomische und juristische Systeme zu schaffen, sind zahlreich; vgl. *Klaus Bosselmann*, Im Namen der Natur. Der Weg zum ökologischen Rechtsstaat, Bern u.a. 1992; *Juan Martinez-Alier* und *Klaus Schlüpmann*, Ecological Economics. Energy, Environment and Society, Oxford 1987; *Jörg Leimbacher*, Die Rechte der Natur, Basel/Frankfurt 1988; *Klaus M. Meyer-Abich*, Wege zum Frieden mit der Natur. Praktische Naturphilosophie für die Umweltpolitik, München 1984.

gültige Normen oder Kriterien bereit. Die Welt als Natur ist vielmehr *„uns und unserem Denken, Handeln und Wünschen gegenüber fremd und gleichgültig“*, und für den Umgang damit gibt es keine Wahrheiten. Erforderlich sind stattdessen *„Streit und Auseinandersetzung von Einzelnen, sozialen Gruppen und Institutionen um die Definition und Durchsetzung dessen, was dem Wohl der Menschen und dem Frieden, auch dem Frieden mit der Natur, dient“*[21].

Mit diesen Fragen haben sich zahlreiche deutschsprachige Untersuchungen zur Umweltgeschichte befasst und dabei einen Schwerpunkt auf die letzten zweihundert Jahre gelegt, d.h. die Zeit seit dem großen Bevölkerungsanstieg und der Industrialisierung. Dabei haben sie vor allem untersucht, welche Veränderungen menschliches Verhalten in den drei Medien – Luft, Wasser und Boden – hervor rief, wie Zeitgenossen darauf reagierten, ob sie Gegenmaßnahmen ergriffen und wie diese aussahen, welche Erfolge sie erzielten oder – weit häufiger – warum so wenig geschah[22]. Anfangs ging es auch bei diesen Arbeiten vielfach darum, nach falschen Weichenstellungen oder Sackgassen[23] zu suchen, doch wie in der allgemeinen Umweltdebatte ist auch bei Umwelthistorikern seitdem eine gewisse Ruhe eingekehrt. Sie untersuchen ihre Themen mittlerweile mit größerer Gelassenheit und stellen nicht nur Krisen fest, sondern auch zahlreiche Bemühungen, etwas gegen diese zu unternehmen und dabei Erfolge zu erzielen[24]. Vermehrt findet zudem die erwähnte Historisierung statt, die u.a. zahlreiche Untersuchungen darüber hervorgebracht hat, warum ab den 1970er Jahren ein derart ausgeprägtes Krisenbewusstsein vorherrschte und wie die erwähnten Schreckensszenarien entstanden[25].

[21] *Willi Oelmüller*, Für eine Ökologiediskussion diesseits von Systemdenken und Moralismus, in: Ökologische Probleme im kulturellen Wandel, hg. v. Hermann Lübbe und Elisabeth Ströker, (Ethik der Wissenschaften, 5) Paderborn 1986, 115.

[22] Als Auswahl genannt seien *Arne Andersen*, Historische Technikfolgenabschätzung am Beispiel des Metallhüttenwesens und der Chemieindustrie, 1850-1933, (Zeitschrift für Unternehmensgeschichte, Beiheft 90) Stuttgart 1996; *Ulrike Gilhaus*, „Schmerzenskinder der Industrie“. Umweltverschmutzung, Umweltpolitik und sozialer Protest im Industriezeitalter in Westfalen 1845-1914, Paderborn 1995; *Jürgen Büschenfeld*, Flüsse und Kloaken. Umweltfragen im Zeitalter der Industrialisierung (1870-1918), Stuttgart 1997; *Franz J. Brüggemeier*, Das unendliche Meer der Lüfte. Industrialisierung, Umweltverschmutzung und Risikodebatten im 19. Jahrhundert, Essen 1996; Umweltbewältigung. Die historische Perspektive, hg. v. *Gerhard Jaritz* und *Vera Winiwarter*, Bielefeld 1994; Umweltgeschichte. Umweltverträgliches Wirtschaften in historischer Perspektive (Geschichte und Gesellschaft, Sonderheft 15), hg. v. *Werner Abelshauser*, Göttingen 1994; weitere Literatur-Hinweise und Forschungsberichte bei *Arne Andersen*, Umweltgeschichte. Forschungsstand und Perspektiven, in: Archiv für Sozialgeschichte 33 (1993), 672-701; *Joachim Radkau*, Was ist Umweltgeschichte?, in: Umweltgeschichte, hg. v. Werner Abelshauser, 11-28; *Franz J. Brüggemeier*, Umweltgeschichte. Erfahrungen, Ergebnisse und Erwartungen, in: Archiv für Sozialgeschichte 43 (2003), 1-18.

[23] So in dem von *mir* und *Thomas Rommelspacher* herausgegebenem Sammelband mit dem Titel: Besiegte Natur. Geschichte der Umwelt im 19. und 20. Jahrhundert, München ²1989.

[24] Einen guten Überblick darüber gibt der Sammelband Umweltgeschichte. Themen und Perspektiven, hg. v. *Wolfram Siemann*, München 2003.

[25] Aktuell dazu mehrere Beiträge in: Natur- und Umweltschutz nach 1945. Konzepte, Konflikte, Kompetenzen, hg. v. *Franz J. Brüggemeier* und *Jens I. Engels*, Frankfurt/M. 2005; ferner: Wird Kassandra heiser? Die Geschichte falscher Ökoalarme, hg. v. *Frank Uekötter* und *Jens Hohensee*, Stuttgart 2004.

2. Disziplin, Methode und Institution

Die vorliegenden Untersuchungen haben sich auf einzelne Medien, besondere Entwicklungen oder Fallbeispiele konzentriert und die geforderten umfassenden ökologischen Fragen und Methoden ausgeklammert, so daß sie anfangs etwas herablassend als „kleine Umweltgeschichte" bezeichnet wurden. Ihre Ergebnisse seien im Einzelnen hilfreich, doch sie könnten die *„größer dimensionierten Fragen* [...] *nicht lösen"*[26]. Das war ein etwas voreiliges Urteil. Fraglos haben diese Arbeiten die aktuelle Umweltkrise nicht gelöst. Dazu ist vielleicht erforderlich, was Lewis Mumford schon vor zwanzig Jahren gefordert hat: *„Um zu ihrer Rettung zu gelangen, wird die Menschheit eine Art spontaner religiöser Bekehrung vollziehen müssen: eine Bekehrung von einem mechanischen Weltbild zu einem organischen"*[27]. Das ist emphatisch formuliert, doch dieser Vorschlag besitzt weder besonders große Erfolgsaussichten, noch ist er gut begründet. Vielmehr hat sich zum einen gezeigt, daß auch ohne die geforderte Bekehrung zu einem neuen Weltbild deutliche Verbesserungen beim Umgang mit der Umwelt zu erzielen sind; und zum anderen wurde deutlich, daß auch weniger ambitionierte Methoden es erlauben, historische Entwicklungen mit Ertrag zu untersuchen und zu analysieren.

Dazu haben Umwelthistoriker/innen auf eine Vielzahl methodischer Zugänge zurückgegriffen und sich bei der Technik- und Rechtsgeschichte ebenso informiert wie bei der Politik- und Institutionengeschichte; sie haben Beiträge der Chemie, Biologie und Forstwissenschaft herangezogen und sich generell auf eine Vielzahl unterschiedlicher Disziplinen gestützt[28]. Doch die so vehement geforderte Interdisziplinarität hat sich überwiegend im Kopf der einzelnen Forscher/innen hergestellt, nur selten hingegen in interdisziplinären Projekten. Zum weitaus größten Teil beruhen Arbeiten zur Umweltgeschichte auf den üblichen historischen Überlieferungen, d.h. vor allem auf Schriftquellen, eher selten ergänzt um Abbildungen, Laborbefunde oder auch Messungen usw. Die fraglos sinnvolle Zusammenarbeit zwischen Geistes- und Sozialwissenschaft auf der einen und Naturwissenschaften auf der anderen Seite ist bisher kaum zu erkennen, wenngleich es dazu Ansätze gibt, so im Graduiertenkolleg ‚Interdisziplinäre Umweltgeschichte' an der Universität Göttingen; einem Projekt Freiburger Historiker und Forstwissenschaftler zum Thema Waldsterben und vor allem in den Untersuchungen von Christian Pfister zur historischen Klimaforschung[29]. Zu nennen sind zudem Beiträge aus der historischen Geographie, der Archäologie,

[26] Rolf P. Sieferle, Perspektiven einer historischen Umweltforschung, in: Fortschritte der Naturzerstörung, hg. v. dems., Frankfurt/M. 1988, 351, 355.

[27] *Lewis Mumford*, Mythos der Maschine. Kultur, Technik und Macht, Frankfurt/M. 1977, 807.

[28] Als frühes Beispiel, Mensch und Umwelt im Mittelalter, hg. v. *Bernd Hermann*, Stuttgart 1986; Umwelt in der Geschichte, hg. v. *dems.*, Göttingen 1989.

[29] *Christian Pfister*, Klimageschichte der Schweiz 1525-1860. Das Klima d. Schweiz von 1525-1860 und seine Bedeutung in der Geschichte von Bevölkerung und Landwirtschaft, Bern u.a. ³1988; *ders.*, Climatic variability in sixteenth century Europe and its social dimension, Dordrecht 1999; *ders.*, Wetternachhersage: 500 Jahre Klimavariationen und Naturkatastrophen (1496-1995), Bern u.a. 1999.

Biologie und vergleichbaren Disziplinen, wo immer wieder einzelne Vertreter dieser Disziplinen zur Umweltgeschichte arbeiten, ohne daß der viel beschworene Durchbruch zu erkennen ist[30].

So arbeitet die Umweltgeschichte vorwiegend nicht nur mit den üblichen Quellen, sondern auch mit den vertrauten Methoden. Eine eigene Methode hat sie nicht hervorgebracht; hier waren anfängliche Erwartungen weit überzogen. Das bedeutet jedoch keinen Nachteil. Die Erfahrung zeigt vielmehr, daß Bemühungen um Interdisziplinarität nur langsam vorankommen und daß schon kleine Schritte einen wichtigen Beitrag bedeuten können. Außerdem hat der Rückgriff auf vertraute Methoden den Vorteil, daß Fragestellungen, Themen und Befunde der Umweltgeschichte für die etablierte Geschichtsschreibung anschlussfähig sind, so daß diese deren Veröffentlichungen zur Kenntnis nimmt und dazu übergeht, sie in ihre Darstellungen zu integrieren.

Diesem Befund entspricht, daß die Umweltgeschichte bisher keine eigenen Institute oder Lehrstühle hervorgebracht hat. Das wäre fraglos wünschenswert, ist in Zeiten knapper Mittel und weniger Stellen jedoch nur schwer zu verwirklichen. Stattdessen haben sich Schwerpunkte umweltgeschichtlicher Arbeiten herausgebildet, zuerst in der Technik- und dann in der Wirtschafts- und Sozialgeschichte. In den letzten Jahren ist eine allmähliche Ausbreitung zu beobachten, nicht nur in der Wissenschaftsgeschichte, sondern auch in einer Vielzahl anderer Teilgebiete. Damit verbunden haben einschlägige Vorhaben und deren Förderung ebenso zugenommen wie die Zahl der Tagungen und Workshops. Doch daneben bestehen immer noch große Defizite. In den Handbüchern und großen Darstellungen etwa zum 19. und 20. Jahrhundert tauchen umwelthistorische Themen nahezu gar nicht auf. Ebenso fehlen eigene Schwerpunkte bei der DFG, der Volkswagenstiftung oder anderen Institutionen der Forschungsförderung.

Weit entwickelt hingegen sind die internationalen Kontakte, die anfangs vor allem zu den USA bestanden, wo die Umweltgeschichte zuerst entstand und am stärksten etabliert ist. Mittlerweile existiert zudem ein europäisches Netzwerk, wesentlich angestoßen und organisiert durch die European Society for Environmental History, die 1999 gegründet wurde, 2007 ihren vierten Kongress in Amsterdam abhalten wird und bei deren Etablierung Historiker/innen aus deutschsprachigen Ländern einen wichtigen Beitrag leisteten und noch leisten[31]. Als zusätzliche Schwerpunkte haben sich in den letzten Jahren Großbritannien, Frankreich und Italien etabliert, sowie jüngst vor allem Mittel- und Osteuropa.

[30] *Hansjörg Küster*, Geschichte der Landschaft in Mitteleuropa. Von der Eiszeit bis zur Gegenwart; München 1995; *ders.*, Geschichte des Waldes, Von der Urzeit bis zur Gegenwart, München 1998; *Rüdiger Glaser*, Klimageschichte Mitteleuropas. 1000 Jahre Wetter, Klima, Katastrophen, Darmstadt 2001; vgl. die Titel in Fußnote 28.

[31] http://adm-websrv3b.sdu.dk/ (22.02.2006).

3. Schwerpunkte und Defizite

Wie bei Historikern üblich, wird auch in der Umweltgeschichte intensiv darüber diskutiert, welche Periodisierungen sinnvoll sind. Unstrittig ist die Unterscheidung zwischen einem solaren bzw. agrarischen und dem fossilen Zeitalter, das mit der Industrialisierung entstand, dieses erst ermöglichte, auf absehbare Zeit vorherrschen wird und unsere Umweltprobleme hervor gerufen hat. Die anfangs verbreitete Annahme, das solare bzw. agrarische Zeitalter habe vergleichbare Probleme nicht gekannt und die Natur schonender behandelt, kann – wie erwähnt – nicht mehr aufrecht erhalten werden.

Ohnehin ist der Begriff ‚solares Zeitalter' sehr pauschal, denn er deckt mehrere Jahrtausende menschlicher Entwicklung von den Jägern und Sammlern bis zur Industrialisierung ab und bezeichnet entsprechend viele und unterschiedliche Gesellschaften, so daß er nur eine grobe Orientierung bietet. Er verweist allerdings darauf, daß diese Gesellschaften elementar auf nachwachsenden Rohstoffen aufbauten, treffend als unterschiedlich entwickelte organische Gesellschaften gekennzeichnet werden können und eigene Grenzen des Wachstums besaßen, die etwa Adam Smith noch als unüberwindbar annahm[32]. Diese organische Grundlage hat erst der Aufstieg fossiler Energien verdrängt, und nachwachsende Rohstoffe spielen in heutigen Industriegesellschaften quantitativ nahezu keine Rolle mehr – solange diese Rohstoffe in genügender Menge vorhanden sind. Deren Ende zeichnet sich jedoch ab und damit der Übergang in ein erneutes solares Zeitalter, dieses Mal jedoch in hoch technisierter Form, da die Hoffnung besteht, durch Solarzellen und vergleichbare Techniken die erforderliche Energiemengen zu erzeugen.

Seit einigen Jahren wird darüber diskutiert, ob innerhalb des industriellen Zeitalters weitere Periodisierungen vorzunehmen sind. Insbesondere Christian Pfister hat argumentiert, daß mit dem Übergang zum preiswerten Erdöl Ende der 1950er Jahre eine zusätzliche wichtige Zäsur stattfand, die er mit dem Begriff des ‚Fünfziger Jahre Syndrom' bezeichnet[33]. Kennzeichnend hierfür seien die Durchsetzung der Konsumgesellschaft; der Übergang zur modernen Landwirtschaft mit ihrem verstärkten Einsatz von Maschinen und Kunstdüngern; billige Transportmöglichkeiten, die zur Zersiedlung der Landschaft, einem intensivierten globalen Handel und der Expansion des Tourismus beitrugen und nicht zuletzt ein Anstieg der Emissionen, unter denen derzeit vor allem der Ausstoß an CO_2 große Probleme verursacht und als Hauptursache der globalen Erwärmung gilt. Während die Bedeutung dieser Entwicklungen anerkannt wird, ist umstritten, ob – wie Pfister argumentiert – diese seit den 1950er Jahren eine Beschleunigung erfuhren und ob dafür die preiswerte Energie tatsächlich von entscheidender Bedeutung war. Zu denken ist an andere Faktoren wie den technischen und wissenschaftlichen Fortschritt, an economies of scale, eine Reduzierung

[32] *Edward A. Wrigley*, Continuity, Chance and Change. The Character of the Industrial Revolution in England, Cambridge 1988.

[33] *Christian Pfister*, Das 1950er Syndrom. Der Weg in die Konsumgesellschaft, Bern u.a. 1995.

von Transaktionskosten oder an die stabileren politischen Verhältnisse nach dem Zweiten Weltkrieg[34].

In enger Beziehung zur These vom ‚Fünfziger Jahre Syndrom', aber unabhängig davon entstanden, steht das Konzept der Risikogesellschaft, die ihrem Urheber Ulrich Beck zufolge gegen 1970 entstand und die klassische Industriegesellschaft ablöste[35]. Bis dahin seien Klassenkonflikte und soziale Ungleichheit prägende Merkmale von Politik und Gesellschaft gewesen. Doch angesichts der wachsenden und mittlerweile globalen Umweltbelastungen besäßen diese Unterschiede keine entscheidende Bedeutung mehr. Die ökologischen Risiken beträfen vielmehr alle, unabhängig von Alter, Klasse, Geschlecht oder Wohnort und hätten eine globale Risikogesellschaft erzeugt. So zumindest die ersten Veröffentlichungen zu dieser These, die Beck mittlerweile deutlich revidiert hat. Denn in den letzten Jahren wurde zunehmend deutlich, daß die Umweltrisiken doch nicht alle gleichermaßen betreffen und daß die Folgen sozialer Ungleichheit weiterhin von zentraler Bedeutung sind[36].

Parallel zur These der Risikogesellschaft ist in den letzten Jahren eine weltweite Debatte unter dem Begriff der Nachhaltigkeit geführt worden. Ursprünglich stammt dieser Begriff aus der deutschen Forstwirtschaft des ausgehenden 18. Jahrhundert und bezeichnet eine Bewirtschaftung von Wäldern, bei der jedes Jahr nur so viel Holz geschlagen wird, wie nachwächst. Hieran knüpft die heutige Debatte an und fragt danach, wie industrielle Gesellschaften so wirtschaften können, daß sie nicht durch ihren Verbrauch an Ressourcen und ihre Umweltbelastungen die eigene Existenzgrundlage vernichten, sondern auf Dauer bestehen können. Dabei geht es in der aktuellen Diskussion nicht nur darum, deren physische Existenz durch geringeren Ressourcenverbrauch und abnehmende Belastungen zu sichern. In den angestrebten nachhaltigen Gesellschaften sollen zudem auch unter politischen und sozialen Gesichtspunkten gleichberechtigte und erstrebenswerte Verhältnisse herrschen, so daß zusätzlich zu einer ökologischen auch eine politische und soziale Nachhaltigkeit angestrebt wird, die zudem global gelten soll[37].

Diese Weiterung über die anfänglichen Umweltaspekte hinaus ist verständlich, hat aber auch dazu geführt, daß der Begriff der Nachhaltigkeit immer weiter ausgedehnt wurde, seine Konturen verlor und heutzutage in endlos vielen Varianten besteht. Dennoch: Die darüber geführten Debatten werden anhalten und werfen für Historiker die Frage auf, inwieweit sie diesen Begriff für ihre eigenen Untersuchun-

[34] *Ders.*, Energie und Umweltbelastung. Zum Stand der Diskussion über das ‚1950er Syndrom', in: Siemann (wie Anm. 24), 61-86.

[35] *Ulrich Beck*, Risikogesellschaft. Auf dem Weg in eine andere Moderne, Frankfurt/M. 1986.

[36] *Ulrich Beck* und *Christoph Lau*, Theorie und Empirie reflexiver Modernisierung. Von der Notwendigkeit und den Schwierigkeiten, einen historischen Gesellschaftswandel innerhalb der Moderne zu beobachten und zu begreifen, in: Soziale Welt 56 (2005), 107-35.

[37] Politik der Nachhaltigkeit. Voraussetzungen, Probleme, Chancen – eine kritische Diskussion, hg. v. *Karl-Werner Brand*, Berlin 2002; Nachhaltigkeit. Bilanz und Ausdruck, hg. v. *Wolf-Dieter Grossmann* u.a., Frankfurt/M. 1999; Nachhaltigkeit in naturwissenschaftlicher und sozialwissenschaftlicher Perspektive, hg. v. *Peter Fritz* u.a., Stuttgart 1995.

gen nutzen können. Dabei geht es weniger darum, dessen Entstehung und Veränderungen seit dem 18. Jahrhundert nachzuzeichnen, was interessant und wichtig genug ist. Ertragreicher könnte die darüber hinaus gehende Frage sein, ob vorindustrielle Gesellschaften nachhaltig waren bzw. wie sich der Begriff für ihre Analyse einsetzen läßt[38].

Ansatzpunkte für ein solches Vorgehen hat es in der Debatte darüber gegeben, ob um 1800 in Deutschland ein Holzmangel herrschte und die damaligen Gesellschaften an Grenzen ihres Wachstums stießen, die erst der Übergang zur Kohle überwinden konnte. Die Auseinandersetzung darüber kann als erste große Kontroverse in der deutschen Umweltgeschichte gelten, die nicht nur die Frage eines tatsächlichen oder vermeintlichen Holzmangels behandelte. Vielmehr warf sie auch wichtige Fragen der Überlieferung, der Quelleninterpretation, der Begrifflichkeiten und des methodischen Vorgehens auf, die deshalb besonders wichtig sind, weil die Umweltgeschichte so sehr durch die aktuelle Umweltdebatte geprägt ist[39]. Angesichts der hier herrschenden Befürchtungen stellt sich verschärft das vertraute Problem historischen Arbeitens, daß wir unvermeidlich von heutigen Wahrnehmungen und einem heutigen Problemverständnis ausgehen und dabei zugleich ein Gespür für jeweilige historische Kontexte, Kenntnisse und Wertmaßstäbe bewahren müssen.

Das gilt auch für den zweiten Schwerpunkt innerhalb der deutschen Umweltgeschichte, die Untersuchung von Emissionen und anderen Belastungen, die durch Industrialisierung, Städtewachstum und den zunehmenden Konsum verursacht wurden – während Arbeiten, die zeitlich davor liegende Phasen behandeln, (noch) eine Minderheit bilden[40]. Da die Belastungen früh auftraten und um 1900 ein erhebliches

[38] Als hervorragender Überblick *Joachim Radkau*, Natur und Macht. Eine Weltgeschichte der Umwelt, München 2000; ferner *Rolf P. Sieferle*, Nachhaltigkeit in universalhistorischer Perspektive, in: Siemann (wie Anm. 24), 39-60.

[39] *Joachim Radkau*, Zur angeblichen Energiekrise des 18. Jahrhunderts: Revisionistische Betrachtungen über die Holznot, in: Vierteljahresschrift für Sozial- und Wirtschaftsgeschichte 73 (1986), 1-37; *Rolf P. Sieferle*, Der unterirdische Wald. Energiekrise und industrielle Revolution, München 1982.

[40] Zusätzlich zur Literatur in Fußnote 22 vgl. Umwelt-Geschichte. Arbeitsfelder. Forschungsansätze, Perspektiven, hg. v. *Sylvia Hahn* und *Reinhold Reith*, Wien 2001; *Norman Fuchsloch*, Sehen, riechen, schmecken und messen als Bestandteile der gutachterlichen und wissenschaftlichen Tätigkeit der Preußischen Landesanstalt für Wasser-, Boden- und Lufthygiene im Bereich der Luftreinhaltung zwischen 1920 und 1960, Freiberg 1999; *Engelbert Schramm*, Im Namen des Kreislaufs. Ideengeschichte der Modelle vom ökologischen Kreislauf, Frankfurt/M. 1997; *Dieter Büker*, Mensch – Kultur – Abwasser. Von der Annehmlichkeit für wenige zur Existenzfrage der Gesellschaft. Der Umgang des Menschen mit Abwässern, Essen 2000; *Thomas Bauer*, Im Bauch der Stadt. Kanalisation und Hygiene in Frankfurt am Main, Frankfurt/M. 1998; *Beate Olmer*, Wasser. Historisch. Zu Bedeutung und Belastung des Umweltmediums im Ruhrgebiet 1870-1930, Frankfurt/M. 1998; *Andreas Dix*, Industrialisierung und Wassernutzung. Eine historisch-geographische Umweltgeschichte der Tuchfabrik Ludwig Müller in Kuchenheim, Köln 1997; *Ralf Henneking*, Chemische Industrie und Umwelt. Konflikte um Umweltbelastungen durch die chemische Industrie am Beispiel der schwerchemischen, Farben- und Düngemittelindustrie der Rheinprovinz (ca. 1800-1914), Stuttgart 1994; *Kerstin Kretschmer*, Braunkohle und Umwelt. Zur Geschichte des nordwestsächsischen Kohlenreviers (1900-1945), Frankfurt/M. 1998; *Rainer Karlsch* und *Harm Schröter*, Strahlende Vergangenheit. Studien zur Geschichte des Uranbergbaus der Wismut, St. Katharinen 1996; *Rita Gudermann*, Morastwelt und Paradies. Ökonomie

Ausmaß erreichten, lag es nahe, von Fehlentwicklungen oder Versäumnissen zu sprechen und das Verhalten der damals Handelnden sehr kritisch zu bewerten. Diese Kritik ist verständlich und in weiten Teilen berechtigt; sie hat jedoch auch dazu geführt, sich zu wenig von heutigen Vorstellungen und Erfahrungen zu lösen und so zu verkürzten Einschätzungen zu kommen.

Eine deutsche Besonderheit bildet die Frage, in welchem Verhältnis Konzepte und Vorstellungen des Natur- und Heimatschutzes zum späteren Nationalsozialismus standen. Diese Frage drängte sich früh auf, weil bereits der anfängliche Natur- und Heimatschutz eine rechts-konservative und völkische Prägung besaß, sich während der Weimarer Republik politisch radikalisierte und durch seine führenden Vertreter die Machtergreifung der Nationalsozialisten begrüßte[41]. Wie in anderen Bereichen haben auch hier die Verantwortlichen nach 1945 ihre Nähe zum Nationalsozialismus geleugnet, während mittlerweile zahlreiche Untersuchungen vorliegen, die enge Verbindungen nachweisen. Dabei haben sich die aktuellen Fragestellungen etwas verschoben. Es geht nicht länger vor allen Dingen darum, Affinitäten zum Nationalsozialismus zu untersuchen, sondern danach zu fragen, inwieweit dieser selbst ‚grüne' Elemente besaß und in wichtigen Teilen vielleicht sogar als die erste erfolgreiche grüne Bewegung in Deutschland bezeichnet werden kann[42].

Hiermit ist die Frage verbunden, inwieweit auch die aktuelle Umwelt- und Naturschutzbewegung Gemeinsamkeiten mit Vorläufern im Kaiserreich und vor allem während des Nationalsozialismus aufweist. Hierzu liegen erste Arbeiten vor, die bald durch derzeit noch laufende Untersuchungen ergänzt werden und Teil eines weiteren Schwerpunktes in der deutschen Umweltgeschichte sind, der auf den Jahrzehnten seit dem Zweiten Weltkrieg liegt. Gefragt wird, ob ab Ende der 1950er Jahre tatsächlich die von Christian Pfister festgestellte Zäsur erfolgte; warum die moderne Umweltbewegung entstand und welche Themen diese vorrangig aufgriff; ob wir heute in einer Risikogesellschaft leben; ob sich in den letzten Jahrzehnten Lösungswege abzeichneten oder wie sich die Wahrnehmungen änderten – um nur einige Aspekte herauszugreifen[43].

und Ökologie in der Landwirtschaft am Beispiel der Meliorationen in Westfalen und Brandenburg (1830-1880), Paderborn u.a. 2000; *Bernd Herrmann* und *Martina Kaup*, Nun blüht es von End zu End all überall. Die Eindeichung des Nieder-Oderbruches 1747-1753, Münster u.a. 1997; *Frank Uekötter*, Von der Rauchplage zur ökologischen Revolution. Eine Geschichte der Luftverschmutzung in Deutschland und den USA 1880-1970, Essen 2003.

41 Vgl. *Andreas Knaut*, Zurück zur Natur. Die Wurzeln der Ökologiebewegung, Weinheim 1993; *Gert Gröning* und *Jochen Wolschke-Bulmahn*, Die Liebe zur Landschaft. Teil III: Der Drang nach Osten, München 1987; Der „Generalplan Ost". Hauptlinien der nationalsozialistischen Planungs- und Vernichtungspolitik, hg. v. *Mechthild Rössler* und *Sabine Schleiermacher*, Berlin 1993; Naturschutz und Nationalsozialismus, hg. v. *Joachim Radkau*, Frankfurt/M. 2003.

42 So bezeichnet der amerikanische Historiker *Milan L. Hauner* die Nationalsozialisten als *„the first radical environmentalists in control of a state"*, in: *ders.*, A German Racial Revolution?, in: Journal of Contemporary History 19 (1984), 669-87, Fußnote 46; How green were the Nazis? Nature, Environment, and Nation in the Third Reich, hg. v. *Franz J. Brüggemeier*, *Mark Cioc* und *Thomas Zeller*, Athens, Ohio 2005.

43 *Brüggemeier* und *Engels* (wie Anm. 25); *Jens Ivo Engels*, Naturpolitik in der Bundesrepublik. Ideenwelt und politische Verhaltensstile in Naturschutz und Umweltbewegung 1950-1980, Paderborn 2006; *Kai Hünemörder*, Die Frühgeschichte der globalen Umweltkrise und die Formierung der deutschen Umweltpolitik (1950-1973), Stuttgart 2004.

Ein weiterer, seit einigen Jahren geradezu aufblühender Forschungszweig stellt die Geschichte von Naturkatastrophen in den Mittelpunkt. In diesem Zusammenhang werden die Naturgewalten als historische ‚Akteure' sichtbar, wenn auch immer im Zusammenwirken mit den jeweiligen gesellschaftlichen Verhältnissen. Die Katastrophen erweisen sich als Indikatoren für Mentalitäten, öffentliche Deutungsvorgänge, soziale Sicherungssysteme, technische Vorkehrungen und Wirtschaftssysteme – wie auch als Agenten des Wandels in diesen Bereichen. Nach einigen Konferenzen zu diesem Thema hat sich jüngst ein von der DFG gefördertes „Wissenschaftliches Netzwerk" von Nachwuchshistorikerinnen und -historikern gebildet, die die ertragreiche Forschungsarbeit zu den Katastrophen epochenübergreifend fortsetzen[44].

Schließlich entwickelt sich derzeit ein neues Forschungsfeld mit engen Verbindungen zur Umweltgeschichte: die Geschichte der Infrastrukturen. Hierbei werden die Nutzung natürlicher Ressourcen und die Veränderung natürlicher Umwelten ebenso thematisiert wie technische Bauwerke und schließlich auch die politischen und sozialen Begleiterscheinungen bzw. Voraussetzungen und Folgen dieser Einrichtungen[45].

Insgesamt bietet sich somit ein erfreuliches Bild. Die deutsche Umweltgeschichte, die seit etwas mehr als zwei Jahrzehnten besteht, hat in dieser Zeit einen großen Aufschwung genommen und sich zugleich von allzu tagespolitisch aufgeladenen normativen Konzepten befreit. Gerade in den letzten Jahren ist die Zahl der Veröffentlichungen und insbesondere der Promotionen deutlich angestiegen. Eine feste institutionelle Verankerung durch eigene Lehrstühle oder gar Institute besteht nicht. Doch diese war ohnehin nicht zu erwarten und ist so lange kein größeres Manko, als umweltgeschichtliche Themen und Fragen in die allgemeine Geschichtsschreibung Eingang finden. Bei Handbüchern oder großen Gesamtdarstellungen ist dies noch nicht gelungen, selbst Überblicke zur Bundesrepublik gehen darauf allenfalls kursorisch ein. Doch zahlreiche Indikatoren weisen darauf hin, daß diese Situation sich bald ändern wird, so daß insgesamt eine positive Einschätzung der kommenden Jahre gerechtfertigt ist. Dies gilt umso mehr, als auch in anderen europäischen Ländern wie Frankreich, Italien oder Großbritannien ein Aufschwung der Umweltgeschichte zu verzeichnen ist und die Erwartung rechtfertigen, daß diese Entwicklungen sich gegenseitig verstärken.

[44] Cities and Catastrophes. Coping with Emergency in European History, hg. v. *Geneviève Massard-Guilbaud*, *Harold Platt* und *Dieter Schott*, Frankfurt/M. 2002;, Naturkatastrophen. Beiträge zu ihrer Deutung, Wahrnehmung und Darstellung in Text und Bild von der Antike bis ins 20. Jahrhundert, hg. v. *Dieter Groh*, *Michael Kempe* und *Franz Mauelshagen*, Tübingen 2003.

[45] Grundlegend *Dirk van Laak*, Infra-Strukturgeschichte, in: Geschichte und Gesellschaft 27 (2001), 367-393; im Jahr 2007 erscheint ein Themenheft der Zeitschrift Saeculum über Infrastrukturen, hg. v. *Jens I. Engels* und *Julia Obertreis*.

Angus J.L. Winchester

Common Land in Upland Britain: Tragic Unsustainability or Utopian Community Resource?

How past societies have managed their resources is one of the central concerns of environmental history. At one level this is the subject matter of traditional geography, the study of the interface between society and the natural environment. But from the historian's perspective a "resource" is essentially a human construct and understanding that interface requires us to explore cultural change and the structures of society. Patterns of resource use ultimately reflect patterns of power and perceptions: resource exploitation is a product of perceptions of environmental potential by those with the power to access the resource in question, and perceptions themselves change. The comparatively steady state of the inherent quality of the land itself (its geology, soils, flora and fauna) is viewed through cultural lenses shaped by economic imperatives, the technology of the time and the institutions of society. To the historian, therefore, cultural shifts and changing patterns of power and authority lie at the heart of the history of resource management.

The environment in which I wish to explore these questions is the rough hill land of upland Britain, which survives today in a semi-natural, unimproved state. The most extensive and wildest of such terrain is in the Highlands of Scotland but my focus is on the large areas of northern England which fall into this category in the English Lake District and the Pennine hills (*Ill. 1*, p. 63). Extensive areas of northern English "fell" and moorland still have the status of common land, that is land which is privately owned but over which third parties have rights of utilisation, enabling them to use and take products from the land. A recent calculation suggests that 8,765 individual commons survive in England and Wales, accounting for 1.37 million acres (over 554,000 ha), of which over 700,000 acres (280,000 ha) is in northern England, but these are only a fragment of the much more extensive tracts which were common before the 19th century[1].

Common land exerts a particular fascination for those who are interested in the management of resources. In a world dominated by concepts of private property, resources held in common are often seen as an anachronism, left-overs from an earlier

[1] *Christopher P. Rodgers*, Environmental Management of Common Land: towards a new legal framework?: Journal of Environmental Law 11, 2 (1999), 231-255 at p. 231.

age. Yet common land of various types survives extensively in many parts of Europe and by its very nature, as land over which no single individual can exercise absolute control, it generates records which enable us to explore the cultural context of the decision-making processes governing resource use and sustainability. Common resources have had a mixed press. Perhaps the most influential view, repeated since Aristotle, has been the assumption that, since individuals will generally put their own personal interest first, common property is unsustainable, as individual greed will destroy or over-exploit it. The "*tragedy of the commons*", as expressed in Garret Hardin's famous paper, is that "*Freedom in a commons brings ruin to all*"[2]. However, Hardin's assumption has been questioned, notably in the work of Elinor Ostrom on contemporary common resources, who has shown that communal decision-making by those using a common resource can achieve sustainability and that the "*tragedy of the commons*" need not be inevitable where strong, self-governing local institutions exist[3].

My aim in this paper is to explore these alternative views of common resources by focusing on common land in the hills of northern England and taking a long view, charting the changing patterns of resource use and management from the 12th century to the 20th. The focus is on the management of grazing rights – the principal economic value of the hill commons to agrarian communities – and the central theme running through this paper is the mechanisms through which livestock numbers were controlled and limited. My particular concern is with patterns of power and authority and their encapsulation in the legal and institutional framework of rights in land, which controlled access to and use of resources. My thesis is that cultural perceptions of the northern English uplands have undergone two major transitions in the past eight hundred years, from seigniorial hunting ground to community grazing land in the late-medieval period, and from local resource to national treasured landscape in the 19th century. These changing perceptions were accompanied by changing patterns of power and were reflected in shifts in the legal and institutional framework to resource use.

It is important to note at the outset that Hardin's conclusion that "*freedom in a commons brings ruin to all*" was based on a false premise, namely that individuals would have unfettered access to a common pasture and could put to graze there as many animals as they wished. Taking the long view of the management of common wastes in upland Britain, it becomes clear that use rights were rarely, if ever, unlimited, even in the medieval period. However, the principles and practices of managing access to the common grazings changed as cultural conceptions of the hill wastes evolved.

[2] *Garret Hardin*, The Tragedy of the Commons: Science 162 (13 Dec. 1968), 1243-1248 (reprinted: Managing the Commons, eds. G. Hardin and J. Baden, San Francisco 1977, 16-30).

[3] *Elinor Ostrom*, Governing the Commons: the evolution of institutions for collective action, Cambridge 1990.

Ill. 1 Great Langdale. The mountainous 'fells' of the English Lake District survive here as unenclosed common land. The narrow tongue of enclosed fields in the valley floor contrasts with the semi-natural environment of the common grazings. (Photo: Angus J.L. Winchester)

1. The Medieval Uplands: hunting forest and private pastures

It is striking that most of the mountain and moorland areas of Britain are described as "forest" or "chase" when they first emerge into the written record during the medieval centuries. In England, much of the Pennine chain, from the forest of Cheviot in the north to the great royal forest of the High Peak in the south, large parts of the Cumbrian fells in the English Lake District and of the North York Moors, the hills of southern Shropshire and the Welsh marches, and the moorlands of Exmoor and Dartmoor in south-west England were all forests or chases[4]. Extensive upland forests are recorded in medieval Wales and Scotland as well[5]. It is important to stress that both "forest" and "chase" were legal, rather than ecological, terms, referring to the

[4] *Angus J.L. Winchester*, Moorland Forests of Medieval England: Society, Landscape and Environment in Upland Britain, eds. I.D. Whyte and A.J.L. Winchester, (Society for Landscape Studies Supplementary Series 2) Birmingham 2005, 21-34.

[5] For Scotland see *John M. Gilbert*, Hunting and Hunting Reserves in Medieval Scotland, Edinburgh 1979, 360-363; for a provisional map of forests in Wales see Forests and Chases of England and Wales c.1500-c.1850, eds. *John Langton* and *Graham Jones*, Oxford 2005, viii.

status of the land in question as a hunting ground belonging either to the king or, more usually in these upland areas, to a lay feudal overlord. In the royal forests (such as Dartmoor and the High Peak) the full apparatus of forest law, foresters and forest courts could be brought to bear to preserve game species for the hunt, but private forests or chases also possessed foresters to uphold the overlord's hunting rights. That, at least, was the legal theory: in practice there is little indication that private forests were actively managed for the chase by the 13th century and, even in the royal forests, hunting formed only one aspect of the forest economy.

Indeed, by the 13th century, most of these private forests and chases were treated as an upland frontier containing vast expanses of grazing land with the potential to yield a substantial income for the lord's coffers. Two strategies were employed. In some forests, which may be termed "closed" forests, the lords retained the hill country in their own hands, establishing demesne stock farms, notably "vaccaries" (*vaccariae*), the cattle ranches which were numerous in the Pennines, or granting upland pastures to monastic houses, which established similar, large-scale stock farms on them. In other upland forests ("open" forests) the lords allowed colonisation by peasant communities in the rapid population expansion of the 12th and 13th centuries. New farms and hamlets were established in the valleys. By the 15th century the distinction between "closed" and "open" forests had faded, as demesne vaccaries were leased to tenants and subdivided into smaller farms. The result was a landscape of hill farming communities, dwelling in small hamlets or scattered farms and rearing livestock which they grazed on the unenclosed hill commons[6].

The driving force in the medieval exploitation of the uplands was thus the lords' desire to maximise their income from their forest pastures, whether directly through the profits of their vaccaries, or indirectly through the rents of their tenants. Hunting appears to have played a minor role but the desire to conserve deer for the chase (pursued only fitfully and only in some forests) prompted a concern on the lords' part to conserve woodland as cover and sustenance for game. Despite this, the indications are that the extent of woodland in the uplands declined significantly across the middle ages, probably more as a result of increasing numbers of cattle and sheep trampling seedlings and ringing bark in the winter, thus killing trees and preventing regeneration, than through active felling. Areas of woodland mentioned in 12th and 13th-century sources or implied by place-names are found to have disappeared by the 15th century; income from pannage (payment for the right to put pigs to grub in woodland in the autumn) dwindled from the 13th century, and there is evidence that surviving woodland was coming to be jealously guarded and protected from grazing animals by c. 1450. The deer (and consequently hunting) came to be restricted to enclosed parks, often very extensive, such as the bishop of Durham's vast park in his forest of Weardale or the parks at Leagram and Radholme in Bowland forest and

[6] *Angus J.L. Winchester*, Hill Farming Landscapes of Medieval Northern England: Landscape – the Richest Historical Record, ed. D. Hooke, (Society for Landscape Studies Supplementary Series 1) Birmingham 2000, 75-84.

Musbury in Rossendale forest, but even these had often been leased for pasture by the 15th century[7].

The power of the lords of the forests and the strategies they adopted to raise income from grazing left a legacy that continued to frame patterns of enclosure and pastoral land use in the post-medieval uplands. The extent of enclosure of hillside pastures, in particular, reflected the contrasting origins of settlement in the medieval centuries. In the "closed" forests, dominated by vaccaries and other seigniorial stock farms, large hillside pastures, often covering several hundred acres on the lower slopes of the hills, were enclosed as cow pastures. This process, which continued into the 17th century, resulted in a landscape of large, shared pastures, held in common between the members of each hamlet community. In the "open" forests, by contrast, peasant colonisation often resulted in narrow fingers of enclosure, restricted to arable land and hay meadows on the valley floors. The hillsides remained open and these rough grazings (like the high moorlands beyond the vaccary cow pastures in the former "closed" forests) provided the summer pasture for the livestock of the hill farming communities[8].

By the 16th century most upland communities enjoyed rights of pasture tantamount to the normal common rights found on common wastes outside the upland forests. Yet, in origin, these rights again reflected the distinctive status of the uplands as private forest. Some – perhaps most – pasture rights in the uplands originated in the "agistment" of livestock in the lord's forest. Agistment, a license to graze for which a payment was made, enabled lords to receive income from livestock brought into their forest from outside. The practice almost certainly predated the permanent settlement of the uplands: place-names and the fragmentary documentary record suggest that the hill grazings were used as summer pasture by communities in the surrounding lowlands in the early-medieval period. Where seasonal settlement associated with transhumance and summer shielings survived into the late-medieval centuries, payments akin to those for (and sometimes referred to as) agistment are recorded. In some places individuals in lowland communities retained ancient grazing rights in the upland forests for which separate payments were made, as in the "grasslands" or "grasshouses" of Copeland forest in the Lake District[9]. Even rights which later appear as normal common rights sometimes originated in agistment. In the Northumberland manor of Allendale, for example, sums paid by tenants in 1608 for common of pasture were described in 1547 as payments for the "*agistment of cattle*" and are probably to be traced back to income from "*agistment in the forest*" recorded in 1422. Similar payments for agistment from tenants in the upland valleys are found

7 *Angus J.L. Winchester*, Landscape and Society in Medieval Cumbria, Edinburgh 1987, 100-107; *Winchester* (note 6), 82-83; *Winchester* (note 4), 24-25; *G.H. Tupling*, The Economic History of Rossendale, (Chetham Society, new series vol. 86) Manchester 1927, 33.

8 *Angus J.L. Winchester*, The Harvest of the Hills: rural life in northern England the Scottish Borders 1400-1700, Edinburgh 2000, 52-73.

9 *Winchester* (note 7), 91; similar "grasslands" in Ennerdale are recorded in Cumbria Record Office (Carlisle), D/Lons/W8/7/1, survey of 1747.

in the forests of the Lake District[10]. Such payments ultimately reflect the private status of the forest wastes of the medieval uplands, to which access could only be gained by licence from the lord.

As peasant communities grew in the valleys the distinctive origin of their pasture rights was forgotten and the hills came to be treated as conventional common land, as will be discussed below. By the 16th century agistment proper was restricted to private agistment grounds, often remote pastures, far from habitation, where lords continued to take income from stock brought in from outside the manor. These last vestiges of the private forests were effectively beyond the control of local communities, access to their resources being controlled directly by the lord or his officers.

The dominant theme in the management of resources in the medieval uplands is thus the power of the individual overlord, whether driven by the desire to conserve game for the chase or, much more usually from the 13th century, by the desire to raise income by exploiting his hill pastures and forest wastes for grazing.

2. The Age of the Manor Courts, c. 1400-1750

The settlement of the valleys penetrating the upland forests resulted in a transition in the structures of resource use and management. By the 16th century, most of the uplands were no longer actively managed as hunting reserves but had become the grazing grounds of settled communities of pastoral farmers. In consequence, the balance of power shifted from lord to tenants and local communities evolved management strategies driven by the needs of livestock farming. The 16th and 17th centuries were the heyday of local communal control of common land.

By the 16th century forest status had been all but forgotten in most upland areas. There were some exceptions, notably the royal forests, where game preservation remained a live issue: the conflicting demands of the preservation of deer and the needs of the farming community dominate the records of the royal forest of Bowland in the 1570s, for example[11]. But in most of the northern English uplands the hill grazings had acquired the legal status of "manorial waste", that is land within the boundaries of a manor which remained in a semi-natural state and had not been appropriated to a single individual. Since the statute of Merton in 1236 rights in the soil of such land were vested in the lord of the manor but his ability to enclose and take profit from manorial waste was severely constrained by the use rights of the tenants of the manor. These use rights, which generally included common of pasture (the right to graze animals on the waste), common of turbary (the right to cut turf and peat) and common of estovers (an assemblage of rights to cut vegetation), effectively ensured that common waste remained unenclosed, held in balance between lord and tenants. The commons formed an integral part of the rural economy, providing the bulk of summer

[10] *Winchester* (note 7), 84; *Winchester* (note 8), 94.
[11] Lancashire Record Office (Preston), DDHCl, box 86. See also *Winchester* (note 6), 83.

pasture for livestock, the principal source of domestic fuel in the form of peat, and essential materials such as sods and heather or bracken thatch for roofing, and bracken as litter for bedding livestock. A key feature of the law underpinning common use rights was that they were attached to a holding of a house or land in the manor and could only be exercised to uphold this "dominant tenement". Common of pasture was thus restricted to providing grazing for livestock from the farms within the manor; common of turbary to providing peat for domestic use and not for sale.

Managing the day-to-day exercise of common rights fell to the manor court, an assembly of the tenants summoned by the lord, at which decisions were taken by a jury drawn from the tenants. The courts were formally seigniorial courts, held to uphold the lord's rights and to punish infringements of his prerogatives, but they probably had a hybrid origin, subsuming early village meetings about which we know hardly anything. A memory of early folk assemblies survived in the terminology of some manor court institutions, such as the terms "byrlaw court" and "burlawman", from the Scandinavian word *byjar-log*, a "law community" or "law district". Few records of manor courts in the uplands survive from before 1450 but it clear that strong bodies of local custom, including byelaws governing the use of common land, existed in many manors by then[12].

The records of the manor courts in their heyday in the 16th and 17th centuries enable us to explore the thinking which underpinned the juries' deliberations[13]. The twin aims were to uphold the lord's privileges and to maintain "good neighbourhood", that is friendly relations within the local community. This the courts did by rehearsing local customary law, supplementing it with new byelaws and orders, and imposing financial penalties against those who breached them. Common land brought "good neighbourhood" into sharp focus, as the place where the tensions between individual self-interest and the common good were played out. Sophisticated suites of byelaws were devised to govern the exercise of common rights and to ensure equitable access to resources. In the case of pasture rights, byelaws governing day-to-day practice included restrictions on the types of livestock which might be grazed (diseased animals, goats and geese being particularly closely controlled); seasonal restrictions, notably on rams in the autumn, in order to control the timing of lambing in spring; requirements on the marking of livestock to show ownership; and the allocation of particular parts of the common to livestock of different types or to the stock of particular commoners as in the designated "outrakes" (routes up which livestock

[12] For manor courts see *Paul D.A. Harvey*, Manorial Records, London 1999, 41-68; *Christopher Harrison*, Manor Courts and the Governance of Tudor England: Communities and Courts in Britain 1150-1900, eds. C. Brooks and M. Lobban, London 1997, 43-59; *Winchester* (note 8), 33-48, which also discusses the term "byrlaw".

[13] The role of the manor courts in the management of common land is discussed at length in *Winchester* (note 8). See also *Robert S. Dilley*, The Cumberland Court Leet and the Use of Common Lands: Transactions of Cumberland & Westmorland Antiquarian & Archaeological Society, new series 67 (1967), 125-151.

were driven to pasture) and sheep "heafs" of the Lake District. In this way the manor courts exercised a communal control, limiting individual freedom in order to achieve the aim of fair and sustainable access to resources.

However, the repeated infringements of these rules governing common rights, so numerous in the pages of the manor court books and verdict sheets of the 16th and 17th centuries, are manifestations of tensions over access to resources and also of increasing pressure on resources in a period of rising population and expansion of livestock numbers. The bonds of "good neighbourhood" could be severely strained in the face of, for example, the growth of commercial livestock farming, with its increasing demand for grazing, or a growing landless population, where every new house required access to fuel resources. In seeking to maintain "good neighbourhood" the courts tended to be inherently conservative, frequently citing "ancient custom" when resolving disputes, seeking a return to the way things had been done in the past, but they also attempted to manage change by issuing new or amended byelaws. Nowhere was the challenge greater than in the management of pasture rights.

On the common pastures, two principles were used to control livestock numbers and hence to provide a framework for sustainable grazing regimes. One was the rule of levancy and couchancy, which stated that a tenant of the manor could graze on the manorial waste in summer no more stock than he could keep across the winter on the produce of his holding of land in the manor. This sought to restrict grazing numbers to the carrying capacity of the farmland on the valley floor and to ensure equitable access to the summer grazings by relating the size of pasture right to the size (or, more strictly, productivity) of a holding. It was a pasture right "without number" but nevertheless limited.

The alternative means of controlling livestock numbers was by "stint", that is by articulating the right in numerical terms. Stints were usually expressed as a right to graze so many horned beasts (the right to graze one beast being termed a "beastgate" or "cattlegate"), and agreed multipliers were used to calculate the right for other types of livestock: typically, five to ten sheep could be grazed for one beastgate, while one horse would be rated at two beastgates, for example. Stinting had long been the norm where the available pasture was too small to accommodate all the stock which might be entitled to graze under the rules of levancy and couchancy. In particular, common rights on the stubble of open arable fields and the aftermath of hay meadows were almost always limited by stint, as were rights in the enclosed cow pastures shared by members of a hamlet community.

What was the significance of these alternative ways of defining pasture rights? And what were their implications for sustainability? First, the two forms of pasture right exhibit a striking geographical distribution, the rule of levancy and couchancy being more common in the "open" forests of the Lake District and north Pennines, where peasant settlement had been the norm, whereas stinting was more common in the vaccary country of the central Pennines, the "closed" forests where the lords had retained closer control during the 12th and 13th centuries. There seems thus to have been an association between stinting and the private pastures of forest areas in which lordly control remained dominant. We have already seen the hints that some pasture rights in the uplands originated in agistment on private forest pastures. Levancy and couchancy was incompatible with agistment (since it limited the grazing right to livestock kept over winter on farms in the manor), while stinting placed a numerical cap on livestock numbers without specifying from whence they should come. Where early references to stinting come from areas of strong forest rights, – such as a description of holdings in Weardale forest in 1438, in which each is assigned a pasture right for a specified number of cattle and sheep – they presumably reflect a lordly desire to limit the size of the tenants' flocks and herds in order to preserve grazing either for beasts of the chase or for demesne livestock. In such cases, stinting may thus be an expression of seigniorial control and a legacy of forest status[14].

Such an interpretation of early stinting differs from the usual explanation for the appearance of stints, which see them as an effective way of controlling livestock numbers in response to pressure on grazing reserves. Certainly there is evidence that it was perceived in this way in the 16th and 17th centuries. A long-running dispute centring on claims of overcharging the common in Longsleddale, Westmorland, was settled temporarily when the Court of Exchequer imposed stinting in 1584 and when it flared up again in the 1630s one of the parties again suggested stinting as a way of resolving the conflict[15].

At first sight stinting would appear to have had numerous advantages over pasture "without number". It allowed livestock numbers to be related directly to the carrying capacity of a particular common; a fixed, numerical limit was presumably easier to police than were grazing rights "without number" and, perhaps most importantly, stinting provided a framework capable of accommodating the realities of stock farming. The rule of levancy and couchancy assumed a simple, static system in which each farm was a self-contained unit, yet it is clear that by 1600, and probably earlier, hill farming practice did not accord with this imagined simplicity. A farmer's land could straddle manorial boundaries; hay was bought in to supplement winter fodder grown on the farm; young livestock were sent out of upland manors in the winter to graze on lowland grass several miles away. None of these practices sat easily with a system which defined the size of pasture right in terms of the number of animals which could be wintered at home on the produce of land in the manor. Stinting, by contrast, avoided these difficulties[16].

So, why did stinting not become more widespread across the centuries when manor court juries devoted so much care and attention to the management of common rights? The courts' tendency to fall back on ancient custom would have acted

14 *Winchester* (note 8), 83-84.
15 *Ibid.*, 83 and sources there cited.
16 Some manor courts explicitly allowed stock to be brought in where an individual did not possess enough animals to make up his full stint: *Robert S. Dilley*, Agricultural Change and Common Land in Cumberland 1700-1850, unpublished PhD thesis, McMaster University 1991, 303; Fountains Abbey Lease Book, ed. *David Michelmore*, (Yorkshire Archaeological Society Record Series 140) Leeds 1981, 309 [Kilnsey, 1534]; Yorkshire Archaeological Society Library (Leeds), DD 121/1/4 (17) [Langstrothdale c. 1589].

as a brake on innovation in this as in other spheres and, moreover, reaching agreement over the carrying capacity of a common and the size of individual commoners' stints would almost certainly have placed even greater strains on "good neighbourhood". That may have been the case where attempts to introduce stinting appear to have failed[17], but there may also have been factors positively favouring pasture rights "without number". Communal self-interest may have promoted flexibility in the interpretation of the rule of levancy and couchancy. Some manors accepted that a limited amount of hay could be bought in; although there were some attempts to legislate against away-wintering of young sheep from Lake District valleys in the later 17th century, we know that the practice was already well established and that some courts explicitly allowed a limited number to be wintered away from home[18]. By allowing hill farmers to break out of the straitjacket of complete dependence on the produce of marginal land in the cold, wet upland valleys, such practices enabled them to increase their livestock numbers. But this, in turn, brought the danger of over-grazing: flexibility for short-term gain might be unsustainable in the longer term.

One of the puzzles surrounding the history of common land in upland England is why the sophisticated systems of control developed by the manor courts collapsed so suddenly in the 18th century. The system had an inherent weakness, namely that, since the manor court was not a self-governing folk meeting but a court summoned by the lord, the holding of courts depended on the lord and an effective manorial administration. Widespread enfranchisement (the conversion of customary tenures into freeholds) reduced the direct influence of the lord in local affairs and marginalised the role of manor courts in estate administration. Once the courts were no longer central to the relationship between lord and tenants, lords were tempted to save expenditure by letting the courts fall into abeyance. Studies of manor courts in Cumbria have demonstrated that the system collapsed quickly from c. 1720: irregular or infrequent court meetings on many manors went hand-in-hand with a decline in the numbers of byelaws, orders and presentments in the court records[19]. By 1800 few manor courts were effective regulators of common resources and this collapse of traditional management systems coincided with a marked shift in the cultural context of common land.

17 For example at Longsleddale (cited above, note 15) and by the manor court at Coniston, Lancashire, in 1796: *Eleanor A. Straughton*, Common Grazing in the Northern Uplands: land, society, governance, since circa 1800, unpublished PhD thesis, University of Lancaster 2005, 150-153.

18 *Winchester* (note 8), 81, 96-97. In 1717 Ennerdale court allowed each tenant to winter up to 40 sheep out of the manor: Cumbria Record Office (Carlisle), D/Lons/W8/7/1, 285.

19 *Charles E. Searle*, Customary Tenants and the Enclosure of the Cumbrian Commons: Northern History 29 (1993), 126-153 at 138-139; *Dilley* (note 16), 547; *Straughton* (note 17), 129a.

3. Upland Commons since the Agricultural Revolution

A second transition in the context of resource management on the upland wastes took place across the 19th century as common land came to be viewed as a national resource and the uplands to be seen as treasured landscapes to be preserved for the common good. This shift in perceptions was intimately connected with "improvement", that sea change in the conception of landed resources which surged from the 17th century, reaching a crescendo in the century of Agricultural Revolution between c. 1760 and c. 1860. The semi-natural state of land was no longer accepted as a given and reclamation and intensification of use became economic and cultural imperatives. Custom and common property were seen as obstacles and were disdained[20]. Central to "improvement" was enclosure, a term which encompassed both a legal and a physical transformation. At its heart lay the extinguishment of common rights and their replacement by the exclusive rights of individuals over defined areas; the consequence was generally the physical division of former unenclosed common land by walls or hedges. An act of Parliament was normally required to effect enclosure, disentangling the rights of lord and tenants in order to enable exclusive ownership over sections of the waste to be allotted to each. Thousands of hectares of the uplands were enclosed by Act of Parliament in the 19th century (*Ill. 2*, p. 72), substantially reducing the acreage of common land in the uplands, but it by no means extinguished it entirely. Common land still accounted for over 25 per cent of the historic county of Westmorland and over 11 per cent each of Cumberland and the North Riding of Yorkshire in the 1960s[21].

The yearning for wild places, which grew in response to urban growth and industrialisation, coincided with the taming of those same wild hillsides through Parliamentary enclosure. By the 1860s the dominant ethos concerning the value of common land had moved from improvement for agriculture to preservation for amenity and recreation. The Commons, Open Spaces and Footpaths Preservation Society was founded in 1865. Statutes, from the Metropolitan Commons Act of 1866 to the commons acts of 1876 and 1899[22], increasingly conceived of common land as public space to be preserved as the lungs of the nation, rather than as a farming resource in need of management. When legislators thought of common land they saw the villages greens and amenity heathlands of southern England, not the vast common grazings of the northern fells. But the focus on public access and recreation both intensified and spread to include the upland commons of northern England across the 20th century, notably in the national park movement, which saw large areas of the northern English uplands designated as National Parks in the 1950s. This coincided with a growing awareness of the fragile ecosystems of the uplands and their susceptibility to damage

20 *Edward P. Thompson*, Customs in Common, London 1993, 126-143.

21 *William G. Hoskins* and *L. Dudley Stamp*, The Common Lands of England and Wales, London 1963, 110.

22 Metropolitan Commons Act 1866: 29 & 30 Vict. c. 122; Commons Act 1876: 39 & 40 Vict. c. 56; Commons Act 1899: 62 & 63 Vict. c. 30.

Ill. 2 The upper reaches of the South Tyne valley in the north Pennines.
The rectilinear enclosures on the valley sides date from the enclosure of former common moorland by Act of Parliament in 1820. (Photo: Angus J.L. Winchester)

through over-grazing, and ecological conservation became another strand in the modern culture of the commons[23].

In the context of these shifting perceptions of common land, how were the resources of the surviving unenclosed commons managed from the early 19th century? This is a comparatively little-studied subject but a recent study of Cumbrian commons by Eleanor Straughton has demonstrated the confusion that has reigned since effective management of the commons by manor courts collapsed[24]. Occasionally the commoners put pressure on the lord to call meetings of the court when problems arose. Sometimes other self-help measures were taken: meetings of commoners or stintholders might be held; parish councils might be used as a local forum to discuss issues affecting common land; or more formal commoners' associations or committees might be established[25]. The problem was that such local, indigenous alternatives to the manor court operated in a legal limbo: stintholders' meetings and commoners'

23 These shifting cultural fashions are explored in *Straughton* (note 17), 33-97.
24 *Straughton* (note 17); *Eleanor A. Straughton*, Beyond Enclosure: upland common land in England and Wales since 1800: Society, Landscape and Environment in Upland Britain, eds. I.D. Whyte and A.J.L. Winchester, (Society for Landscape Studies Supplementary Series 2) Birmingham 2005, 89-98.
25 *Straughton* (note 17), 153-8, 167-92.

associations lacked the legal authority of manor court juries to enforce binding byelaws[26]. Informal solutions seem to have characterised the management of upland commons through much of the 19th and early 20th centuries, made easier, perhaps, by the decrease in the numbers of commoners on many commons as a result of the gathering pace of farm amalgamations. However, local solutions were devised in the context of increasing central government concern over common land, culminating in the Commons Registration Act of 1965[27]. Since then, local management (or lack of it) has taken place in the context of the 1965 Act, which required commoners to register their rights but did not impose management solutions.

The absence of clear management systems since 1800 coincided with – and probably contributed to – growing pressure of livestock numbers on the common grazings. As we have seen, accusations of over-charging the common stretch back to the heyday of manor court control in the 16th century. Complaints of persistent overstocking of common pastures were frequent in Cumbria as the manor courts collapsed across the 18th century and overstocking was of widespread concern by the 19th century[28]. To cite just one example, the tithe commissioner surveying Eskdale and Wasdale, two valleys deep in the Lake District with extensive mountain commons, in 1839 thought that the commons were insufficient to carry the present stock of sheep (*"probably 20 thousand of them"*) and feared that the *"cupidity of each occupier to get as much benefit as he can off the common"* would inevitably lead to increased numbers and a deterioration in the quality of the flocks[29]. Although stock numbers were subject to significant short-term fluctuations and, indeed, fell in the 1920s and 1930s, considerable pressure on surviving common grazings built up after the Second World War. The cumulative result was that the number of sheep in upland counties of England and Wales increased by nearly 62 per cent between 1889 and 1973[30]. Such statistics suggest that the collapse of the manor courts led to a Hardinesque free-for-all, encouraged latterly by hill farming subsidies and headage payments for livestock under the Common Agricultural Policy.

The collapse of traditional management systems, accompanied as it seems to have been by persistent problems of overstocking, thus paved the way for a drift towards anarchy on those commons which survived the wave of Parliamentary enclosure. One strategy for recovering control was a greater use of stinting and there was a considerable increase in the number of stinted commons and pastures in upland northern England in the 19th century. Some were a product of Parliamentary enclosure itself, when awards created "regulated" stinted pastures on sections of the waste deemed too poor to be divided into individual allotments. The process was encouraged by the General Inclosure Act of 1845 and resulted in the conversion of manorial waste into

[26] *Rodgers* (note 1), 239.
[27] Commons Registration Act 1965, c. 64.
[28] *Searle* (note 19), 135-7; *Straughton* (note 17), 148, 227.
[29] The National Archives, Public Record Office (Kew), IR 18/716, report by J.J. Rawlinson, question 11.
[30] *David Grigg*, English Agriculture: an historical perspective, Oxford 1989, Table 18.1.

jointly owned pastures regulated by stint[31]. Further opportunities to convert to stinting followed the Commons Act of 1876, which enabled whole commons to be regulated, but only seven upland commons took advantage of the act. The Act's insistence that public access and recreation should accompany regulation of a common and the Home Office's reluctance to approve byelaws controlling grazing limited its usefulness to commoners[32].

The move towards stinting was therefore far from complete. In 1963 it was estimated that more common land (46 per cent of the total) in England and Wales was used as unstinted grazing than was used as stinted grazing land (33 per cent of the total[33]. But the language of stinting – the assumption that a common right could always be expressed in numerical terms – had gained an ascendancy by the middle of the 20th century. Indeed, the Commons Registration Act of 1965 required pasture rights to be registered numerically, even on commons which had continued to be governed by the rule of levancy and couchancy. The result was not true stinting: there was no attempt to relate the numbers claimed to the carrying capacity of the common. Moreover, by articulating pasture rights in numerical terms, the Act led to an intensification of a problem to which numerical stints were often subject, namely the severance of the right from the land to which it had been attached. Unlike a common right governed by the rule of levancy and couchancy, a stint could be viewed as a free-standing piece of property which could be traded. The commons of Buttermere in the heart of the Lake District (*Ill. 3*) illustrate the confusion and unforeseen consequences caused by the 1965 Act. First, rights appear to have been registered more than once, or numbers to have been grossly inflated in the process of expressing a right "without number" in numerical terms. On Buttermere's mountainous common grazings, pasture rights were registered for 21,000 sheep, yet the carrying capacity of the common has been estimated at fewer than 9,000. Then, since registration, a "*significant proportion*" of these registered rights have been sold or leased, sometimes to people living far from the common[34]. So great is the concern over the sale of stints that the Commons Bill 2006 forbids the severance of common rights from the property to which they are attached[35].

The history of the hill commons of northern England since the collapse of manor court control has been characterised by opportunism, conflicting interests and sticking plaster solutions.

31 As, for example, at Castlerigg (1849) and Bootle (1857) fells in the Lake District and Kirkland fell (1866) in the north Pennines: *Straughton* (note 17), 195a.

32 The upland commons regulated under the 1876 Act were: Matterdale in the Lake District; East Stainmore, Crosby Garrett and Winton & Kaber in eastern Westmorland; and Abbotside, Beamsley and Langbar in the Yorkshire Pennines: *Straughton* (note 17), 208-21.

33 *Hoskins* and *Stamp* (note 21), 109.

34 *Stroughton* (note 17), 104.

35 Commons Bill 2006, section 9, subsections 5, 14.

Ill. 3 Buttermere, a tourist honeypot in the heart of the English Lake District. Surviving common land on the 'fells' continues to be grazed by sheep but is also valuable amenity land, popular with hill walkers. (Photo: Angus J.L. Winchester).

4. Reflections on the environmental history of common land

Most of this paper has been concerned with the comparatively straightforward task of tracing the changing framework of resource management on the upland commons across the centuries. Much more challenging is the second theme of this session of the conference, that much-used term "sustainability". The concept of sustainability – or, rather, of unsustainability – is central to much of the discussion of common resources and was, of course, the central theme of Hardin's "*Tragedy of the Commons*". Taking the long view, we must ask, first, whether sustainability would have been recognised as an aim by those managing common land in the past and, if so, what was being sustained and for whom? In each of the three phases of history discussed above we find an awareness of tension between the conflicting demands of different resource uses: game preservation *versus* raising income from pasture in the medieval centuries; individual self-interest *versus* "good neighbourhood" since the 16th century; amenity value *versus* agricultural value of common lands since the 19th. Sustainability may be rarely stated explicitly, but an awareness of the potential loss of a resource and a desire to take action to preserve it run through the records.

Conceptions of what ought to be preserved have changed as power structures and perceptions of the uplands as a resource have changed. In the medieval period the dominant aim appears to have been to maximise the lord's income through establishing stock farms and selling grazing through agistment. Though there is little documentary evidence for local conditions before the 15th century, it seems likely that population levels and livestock numbers would have been low in comparison with modern times and that pressure on grazing resources may not have been great. But the centuries of colonisation put pressure on the older conception of the uplands as hunting forest, as livestock competed with game.

In the early-modern period the driving force underlying the byelaws promulgated by the manor courts was "good neighbourhood" – the maintenance of friendly relations within the community here and now – rather than a desire to conserve resources for future generations. However, a forward-looking concern to conserve is visible in some manor court rulings, such as the byelaws aimed at limiting the environmental damage caused by peat digging and turf stripping, which required commoners to drain the water and replace the sod after digging peat, so that surface vegetation could grow again, and to refrain from cutting turf where livestock would gather or be fed in order to prevent erosion[36]. Sustainability was implicit in the management of common grazings. It was obvious to any stock farmer that livestock did not thrive if pastures were over-grazed and the notion that a common could only bear a finite number of animals was explicit in stinting and implicit in the rule of levancy and couchancy. A desire to sustain the quality of the flocks and herds grazing on the commons can be traced back to the 16th century, even if persistent and recurring complaints of overstocking suggest that success in achieving this was often limited.

In drawing together this attempt to survey the long-term history of resource management on the commons in Britain, the rather gloomy conclusion that sustainability was rarely achieved should be tempered by recognising the positive aspects of manor court control. It can be argued that the regulation of common rights by manor court byelaws constituted a rounded and sustained attempt at resource management, contrasting starkly with both the "asset stripping" approach of medieval lords and the confusion and near anarchy of the 19th and 20th centuries. The manor court system, self-determining and locally based, with effective mechanisms for monitoring compliance with byelaws, imposing sanctions and resolving disputes, fulfilled many of the requirements identified by Elinor Ostrom as necessary for enduring and sustainable management of common resources[37]. Moreover, the manor court juries were guided by a principle beyond economic imperatives and the maximising of individual profit, namely "good neighbourhood". Just as the "good neighbourhood" imposed restrictions on the individual freedom of members of the manorial community, so the constraints of sustainability need to be accepted, now as in the past, if *that* ideal is to be realised.

36 *Winchester* (note 8), 132.

37 *Ostrom* (note 3), 88-102, 178-81. The editors of the CORN volume on the management of common land in north-west Europe concluded that "*It is difficult* [...] *to find many regions in Europe that did not fulfill most of the criteria laid down by Ostrom for long-enduring, self-managed commons*", The Management of Common Land in North-west Europe, c. 1500-1850, eds. *Martina De Moor, Leigh Shaw-Taylor* and *Paul Warde*, (CORN Publication Series 8) Turnhout 2002, 251.

Bill Luckin

Unending Debate: Town, Country and the Construction of the Rural in England, 1870-2000

I

The green belt continues to retain its potency as a symbol of a deep gulf between town and country in modern and contemporary Britain, a buffer pointing to embedded and radically divergent conceptions of differences between the rural and urban spheres[1]. Its very existence underlines the fact that, in the twentieth century and in our own times, the assumed integrity of village life and mores have counted for significantly more than celebration of regional or civic values and culture.

The belt was originally intended to fence off agricultural land bordering the metropolitan perimeter from ever more rapid, and as many contemporaries termed it, octopus-like extra-urban expansion[2]. Reaching back to the initiatives of the London County Council and the Campaign for the Preservation (later Protection) of Rural England in the early 1930s, the belt was strengthened by the post-war Greater London Plan, masterminded by the extraordinarily influential planning supremo, Patrick Abercrombie[3]. In 1955 and again in 1971 it was finally and formally installed as an instrument of central and local bureaucratic control[4]. The ground had been prepared in two phases, between the 1870s and the beginning of the First World War, and between 1918 and the resumption of hostilities in 1939. Thereafter, the period of reconstruction between the mid-1940s and the late 1950s witnessed increased commitment on the part of government and a now mature planning bureaucracy to the idea that one or another variant of an essential rural England must be insulated against urban and suburban development[5]. This orthodoxy remained dominant for 40 years. Only in the recent past, with complex questions being asked about changing relationships between

1 The major conceptual issues are spelt out in *M.J. Elson*, Greenbelts: Conflict Mediation in the Urban Fringe, London 1986.

2 See the seminal interwar preservationist polemic, *Clough Williams Ellis*, England and the Octopus, London 1928. A decade later, Ellis edited a companion volume on the same theme and called it Britain and the Beast, London 1938.

3 *Gerald Dix*, Patrick Abercrombie 1879-1957: Pioneers in British Planning, ed. Gordon E. Cherry, London 1981, 103-30.

4 *Peter Hall et al*, The Containment of Urban England, vol 2: The Planning System, London 1973, 73.

5 *Peter Hall*, Cities of Tomorrow, An Intellectual History of Urban Planning and Design in the Twentieth Century, Oxford 1996, 185 and 333.

town, suburbia and countryside, has there been a weakening of the idea that rural England must remain indefinitely sacrosanct[6]. During the 1990s and on into the beginning of the new century, this reevaluation gave rise to the founding of the Countryside Alliance. The Alliance capitalized on a wide range of rural discontent, including closure of village primary schools, banks and post offices, the adverse impact of the Common Agricultural Policy on farmers attempting to make a living from small acreages, the trimming or eradication of rail and bus services and governmental commitment to the control or abolition of hunting and coursing[7]. The Alliance has revived the venerable idea that modern Britain is dominated by London, and administered predominantly in the interests of the inhabitants of London and other great conurbations.

Between the 1870s and 1945, and indeed even earlier, anti-urbanism and anti-metropolitanism played a prominent role in British, and even more so in English, cultural life. Since the medieval period the precocious demographic and spatial growth of the capital had engendered widespread apprehension and hostility. This generated elite and popular literary traditions associating London with corruption, environmental and moral pollution and disdain for the assumed ignorance of the inhabitants of smaller towns and villages[8]. During the hundred and thirty years with which this essay is concerned, diffuse consensus was transformed into a mature anti-urban, anti-metropolitan discourse articulated by organizations such as the Kyrle Society, the Royal Society for the Protection of Birds, the Society for Saving the Natural Beauty of the Lake District, the Council for the Preservation of Rural England, and later and in our own times, the National Trust and English Heritage.

Associated literary lineages have been long and distinguished, running from Chaucer to Spencer, from Shakespeare and Jonson to Milton's *Comus*, and, in the eighteenth and nineteenth centuries, from Gay, Goldsmith, Clare and Blake to the pioneering laureate-cum-proto-ecologist, William Wordsworth[9]. Thereafter, representative figures included Ruskin and Morris, Gissing, Conrad, Woolf, Eliot, Lawrence and Forster. In their different ways, these thinkers and writers massively idealized the rural sphere, and castigated urban-industrialism as a progenitor of soulless modernity and interpersonal alienation. Simultaneously many writers and cultural critics excoriated the "*civilization of the machine*"[10]. Anti-urban and anti-metropolitan

6 *M.J. Elson et al*, Green Belts and Affordable Housing. Can We Have Both?, Bristol 1996.

7 The aims of the organization are succinctly described at http://www.countryside-alliance.org.

8 See, for example, *Lawrence Stone*, The Crisis of the Aristocracy, Oxford 1965, 386-96; *R. Williams*, The Country and the City, London 1973 and *P.J. Corfield*, The Impact of English Towns 1700-1800, Oxford 1982, ch. 8.

9 See, in particular, *Frank Kermode*, Shakespeare's Language, London 2000, 78-81; *Max Byrd*, London Transformed: Images of the City in the Eighteenth Century, New Haven and London 1978; *Jonathan Bate*, Romantic Ecology: Wordsworth and the Environmental Tradition, London 1991 and *Alan Robinson*, Imagining London, 1770-1900, London 2005.

10 See, for example, *D.H. Lawrence*, Nottingham and the Mining Countryside: Phoenix. The Posthumous Papers of D.H. Lawrence, ed. E.D. McDonald, London 1936, 133-40 and *E.M. Forster*, The Machine Stops: The Collected Short Stories, London 1947, 115-50. The same author's Howard's End, London 1910, is overwhelmingly preoccupied with the social and relational implications of urban-technological change.

writers, particularly Gissing, Morris and Richard Jefferies, created rural utopias as a counterweight to town-based addiction to mass production and mass consumption. Eliot, echoing ideas first articulated by Blake in his terrifying poem, *London*, and by Wordsworth in the seventh book of the *Prelude*, mordantly associated urban gigantism with literary analogues to Durkheimian anomie[11]. Among composers Elgar, Delius, Vaughn Williams, Finze, Ireland, Britten and the early Tippett created soundscapes which reflected similar anti-urban preoccupations[12]. English painters from the pre-Raphaelites to Spencer, Grant, Nash and Sutherland remained at least partially in thrall to the rural idyll[13].

Above all, however, it was London – and the condition of London – which obsessed anti-urbanists in the long period between the fifteenth and the mid-twentieth century. The premature demographic expansion of the capital, its all-embracing incorporation of the super-wealthy and the abjectly poor, and its seemingly illimitable spatial ambition, promised to undermine the sacred project of civilizing the city. The capital boasted a population of a quarter of a million at the beginning of the seventeenth century, three times that number a century later, a million a century after that, 2.5 million in 1850, 4.5 million in the 1880s and no fewer than seven million at the end of the First World War[14]. From the 1870s onwards the populations of the largest of the metropolitan townships, or registration districts, spiralled upwards to between 150,000 and 300,000. Had they possessed independent urban status, places like St Pancras, Marylebone, Wandsworth, Camberwell and Hammersmith would have occupied an impressive position in the league tables assiduously compiled by mid- and late Victorian and Edwardian social statisticians and reformers[15]. Several substantial "new industrial districts" in the midlands and north – Nottingham, Leicester, Derby, Bolton and Blackburn – were significantly smaller than the largest of the London suburbs, a point lost on commentators more concerned with the alleged failure of the capital to modernize and democratise its supposedly medieval governmental system than with comparative urban demographics[16]. For their part, most Londoners appeared to know and care little about life in urban-industrial Britain.

11 *Robert Crawford*, The Savage and the City in the Work of T.S. Eliot, Oxford 1987.

12 See in particular *Wilfrid Mellers*, Vaughan Williams and the Vision of Albion, London 1989.

13 *Michael Rosenthal*, British Landscape Painting, Oxford 1982, 124-78.

14 On the long-term demographic development of London see *E.A. Wrigley*, A Simple Model of London's Importance in Changing English Society and Economy 1650-1750: Past and Present 37 (1967), 44-70; *P.J. Waller*, Town, City and Nation: England 1850-1914, Oxford 1983, 24-9 and *P. Garside*, London and the Home Counties: The Cambridge Social History of Britain 1750-1950, vol 1: Regions and Communities, ed. F.M.L. Thompson, Cambridge 1990, 498-501.

15 The pre-eminent example is *Adna Ferrin Weber*, The Growth of Cities in the Nineteenth Century: A Study in Statistics, New York 1899.

16 *D. Owen*, The Government of Victorian London 1855-1889: The Metropolitan Board of Works, the Vestries and the City Corporation, ed. Roy MacLeod, Cambridge MA 1982; *J. Davis*, Reforming London: The London Government Problem, Oxford 1988 and *Bill Luckin*, The Metropolitan and the Municipal: The Politics of Health and Environment in London, 1860-1920: Cities of Ideas. Civil Society and Urban Governance in Britain 1800-2000. Essays in Honour of David Reeder, eds. R. Colls and R. Rodger, Aldershot 2004, 46-66.

Nevertheless, and despite the existence of deeply ingrained anti-metropolitanism in places like Birmingham, Manchester and Leeds, the sheer longevity and scale of the capital ensured that London became synonymous with cities and the ills of cities in general. The East End symbolized hopeless poverty, and the metropolitan suburbs provided a paradigm for the way in which major centres of population preyed upon and appropriated hitherto administratively independent extra-urban communities. London, a giant among giants, was perceived as a constant threat to the spatial integrity of small towns and villages lying just outside what would later come to be designated the greater metropolitan region. Little wonder, then, that social and cultural critics, and, later, planners and environmentalists concentrated on the problems of the London-dominated south, rather than the midlands and north, the arable English heartland rather than Scotland, Ireland or Wales. Together with a cluster of organizations concerned with the protection of the Lake District, it was metropolitan intellectuals, bureaucrats and politicians who shaped rural preservationism in the later nineteenth and early twentieth centuries. In that sense, as in many other things, Anglocentric prevailed over larger British values and concerns[17].

II

Urban reformers, disturbed by the overweening outward expansion of the capital settled to a quite astonishing degree on a single, phantasmagoric solution – stabilization of metropolitan population through a recolonization of traditional village communities[18]. Theories of this kind were predicated on rejuvenation of rural trades, crafts and industries believed to have flourished before the onset of capitalistic farming in the post-medieval period. "Reskilled" unemployed dockers would now enjoy healthier and happier lives in village and hamlet. Multiform social, economic and epidemiological problems associated with the inner city would be partially solved, and the quality of the national stock, central to intellectuals and reformers in thrall to the idea of urban degeneration and eugenics, significantly reduced[19]. Aggregate metropolitan population and movement from inner core to suburban periphery would be held steady, giving rise to a deceleration in the rate of

[17] Here environmental history interacts with larger historiographical debates about shifting conceptions of England and Britain. See *Catherine Hills*, The Origins of the English, London 2003; *Norman Davies*, The Isles: A History, London 2000; *Krishan Kumar*, A History of English National Identity, Cambridge 2003 and *Julia Stapleton*, Sir Arthur Bryant and National History in Twentieth Century Britain, Lanham MD 2005.

[18] See, in particular, *Jan Marsh*, Back to the Land: The Pastoral Impulse in England 1880-1914, London 1982; *P.C. Gould*, Early Green Politics: Back to Nature, Back to the Land and Socialism in Britain, 1880-1900, Brighton 1988 and *D. Hardy*, Alternative Communities in Nineteenth Century England, London 1979.

[19] For comprehensive bibliographical guidance on these much debated issues see *D. Stone*, Breeding Supermen: Nietzsche, Race and Eugenics in Interwar Britain, Liverpool 2002 and *Mike Hawkins*, Social Darwinism in Europe and America 1860-1945: Nature as Model and Nature as Threat, Cambridge 1997.

suburban expansion, and capping "colonising" pressure on neighbouring extra-metropolitan communities.

This panacea, combining Ruskin and Morris's rural utopianism with what would have amounted to semi-compulsory relocation policies, ignored what Karl Marx described as the real existing conditions of demand and supply in urban and rural labour-markets. Its emergence and discursive dominance coincided with a period of agricultural recession. The idea of a nationwide "great depression" between the early 1870s and mid-1890s has been banished from the literature. However, there is now agreement that regional and climatic factors, together with new and cheaper supplies of meat and cereals from Australasia and Canada, did indeed trigger extremely severe regional difficulties[20]. Moreover, the push and pull of internal migration had already reduced the proportion of the adult population able to derive a living from working on the land. Comprising nearly a fifth of the workforce in the 1850s, agricultural labourers accounted for slightly less than a tenth at the beginning of the twentieth century, a figure that would plunge downwards to little more than a fiftieth over the next 75 years. Little chance, then, of an East Ham docker finding a job in Thomas Hardy's depopulated Dorset or the broad acres of a Norfolk now registering significant levels of emigration to precisely those parts of the Empire – Canada and Australia – that had played a crucial role in increasing arable unemployment and under-employment[21].

Florid rhetoric advocating rural regeneration and relocation of the metropolitan poor to the depths of the countryside appealed both to paternalistic conservatives and modernising progressives[22]. By the final 30 years of the nineteenth century each of these groups had adopted an increasingly negative or ambivalent attitude towards the possibility of creating what would now be termed an ecologically and socially sustainable urban civilisation. A generation earlier, liberals had seen newly emerging self-governing cities as beacons of progress which would play a central historical role in the emancipation of the working-classes and the final destruction of "*feudalism*"[23]. For its part, the landed interest depicted both London and new industrial towns in the midlands and north as inimical to health, happiness and morality. By the 1870s and 1880s both groups had experienced radical ideological and cultural fragmentation and reconstitution: each had been reshaped within the arenas of parliamentary, municipal and pressure-group politics. However, many of those still definable as lib-

[20] For a penetrating overview see *E.J.T. Collins*, Rural and Agricultural Change, ed. Joan Thirsk, (The Agrarian History of England and Wales, vol 7, part 1: 1850-1914) Cambridge 2000, 138-207.

[21] *Dudley Baines*, Migration and the Mature Economy: Emigration and Internal Migration in England and Wales 1861-1900, Cambridge 1985 and *Charlotte Erickson*, Invisible Immigrants: The Adaptation of English and Scottish Immigrants in Nineteenth Century America, London 1972.

[22] Ideological and party political structures during this period are expertly summarized by *K. Theodore Hoppen*, The Mid-Victorian Generation 1846-1886, Oxford 1998, 131-40 and 638-54 and *G.R. Searle*, A New England: Peace and War 1886-1918, Oxford 2004, ch. 4.

[23] *A. Lees*, Cities Perceived: Urban Society in European and American Thought 1820-1940, Manchester 1985, part 1.

erals or progressives now perceived the city, industrialization and urban-industrialism as inadequate to the tasks of reducing economic scarcity, securing individual happiness and creating the good society. For their part, paternalist conservatives claimed that their original doubts about the moral defensibility and spatial viability of the city and manufacturing industry had been well-founded.

At the same time, history and the professionalization of history as an academic discipline, reinforced doubts about the past achievements and future potential of a machine-dominated society. There had, of course, been profoundly pessimistic and critical voices in the early and mid-nineteenth century. Then in 1881-2 Arnold Toynbee radically and authoritatively modified over-sanguine views of the seismic shift towards an industrial culture[24]. This interpretation shaped English social, though not economic, history throughout the Edwardian and interwar eras and beyond. John and Barbara Hammond gained a massive academic and non-academic readership for their studies of town and village labourers during the early phases of industrialization[25]. G.M. Trevelyan's expansive social history, written for a general audience, abounds with references to the cruelty, profiteering and pollution of factory culture, and massively idealizes conditions of life in preindustrial society[26]. This pessimistic *mentalite* lay at the heart of E.P. Thompson's classic *The Making of the English Class*, published in 1963, and sustained itself in residual form until the demise of institutional and academic Marxism in the late 1980s[27].

The Hammonds' catastrophist interpretation was echoed by inter-war intellectuals, rural preservationists and planners. However, it was now also widely believed that knowledge of and exposure to the beauties of nature might do something to bind up the wounds of an urban working-class, said to have been grievously harmed by exposure to the miseries of urban-industrialism. This analysis was applied both to the period of the industrial revolution and to the collapse of the primary-producing manufacturing sector in the 1930s. Thinkers as influential and culturally diverse as Trevelyan – co-founder of the Council for the Preservation of Rural England as well as eminent historian – Abercrombie, J.M. Keynes and E.M. Forster subscribed to one or another variant of this view[28].

At an extreme, D.H. Lawrence insisted on nothing less than an absolute renunciation of the city and all its works – urban-industrialism, the machine, mass consump-

[24] The Study of Economic History: Collected Inaugural Lectures, 1893-1970, ed *N.G. Harte*, London 1971, xx. See, also, *D. Cannadine*, The Present and the Past in the English Industrial Revolution: Past and Present 103 (1984), 131-72 and *Pat Hudson*, The Industrial Revolution, London 1992.

[25] The key works are *J.L.* and *B. Hammond*, The Town Labourer 1760-1832, London 1917, The Rise of Modern Industry, London 1927 and The Village Labourer, London 1928. For incisive historiographical analysis see *R.M. Hartwell*, The Rise of Modern Industry, in his The Industrial Revolution and Economic Growth, London 1971, 81-108.

[26] *G.M. Trevelyan*, English Social History: A Survey of Six Centuries: Chaucer to Queen Victoria, London 1942.

[27] *E.P. Thompson*, The Making of the English Working Class, London 1963.

[28] Each of these writers contributed to *Clough Williams Ellis* (note 2).

tion and mass entertainment[29]. Trevelyan contended that a solution might lie in controlled access to nature. According to this schema, the ills of urban-industrialism would gradually be corrected through popular rambling, youth hostelling and climbing[30]. Patrick Abercrombie adopted a different position, and one which would later underpin dominant values informing post-war reconstruction. Arguing that it might not be possible to undo the miseries of urban-industrialism by mass exposure to nature *in situ*, he urged that the countryside and countryside values must themselves be imported into the city. Were every town-dweller to be given an opportunity creatively to enjoy *rus in urbe*, while at the same time learning about and being educated in the customs, codes and responsibilities of the wild, he or she would become a physically, culturally and morally improved individual[31]. The values of the early twentieth century garden city movement clearly and powerfully underscored Abercrombie's vision[32]. There need be no contradiction between preserving rural values, securing enhanced access to the wilds, or to nature in the city, rebuilding and beautifying urban centres besmirched by dilapidated factory and slum, and championing one or another form of architectural modernism[33]. Each of these positions may be traced back to the perceived metropolitan crisis of the late nineteenth century. Yet, and this point cannot be too strongly emphasized, London was in reality the *least* classically industrial of the great cities of Britain. However, its status as massive and long established super-metropolis ensured that it became a symbol and paradigm to which consensually negative attitudes towards urbanism, industrialism and urban-industrialism were unthinkingly, routinely and irrationally attached.

The pro-rural, anti-urban, anti-metropolitan discourse which flourished in Britain between the 1870s and 1945 and beyond was also rooted in and reinforced by differential response to world war and total war. The period between 1918 and 1939 witnessed an outpouring of literature which lamented national and familial loss or sought to create a compensatory arcadian moral universe. The fiction of Ford Maddox Ford – particularly the magnificent *Parade's End*; P.G. Wodehouse's massively popular evocations of a comically dysfunctional aristocratic and semi-aristocratic ru-

29 There were similarities between Lawrence's position and the even more extreme stance adopted by *Oswald Spengler* in his astonishingly vitriolic denunciation of the city, The Decline of the West, London 1932. On the larger issue of elite attitudes towards the proper ends of leisure during this period see *D.L. LeMahieu*, A Culture for Democracy. Mass Communication and the Cultivated Mind in Britain between the Wars, Oxford 1988, 103-120 and *John Carey*, The Intellectuals and the Masses. Pride and Prejudice among the Literary Intelligentsia 1880-1939, London 1992. See, also, *H. Caygill*, Walter Benjamin. The Colour of Experience, London 1998, ch. 3.

30 *G.M. Trevelyan*, Must England's Beauty Perish? A Plea on Behalf of the National Trust for Places of Historic Interest and Natural Beauty, London 1929 and "Amenities and the State", Clough Williams Ellis (note 2), 183-6. For a scintillating account of the great historian-preservationist's intellectual development see *D. Cannadine*, G.M. Trevelyan: A Life in History, London 1997.

31 *P. Abercrombie*, Town and Country Planning, London 1933 and Planning in Town and Country. Difficulties and Possibilities, Liverpool and London 1937.

32 See *D. Hardy*, From Garden Cities to New Towns. Campaigning for Town and Country Planning 1899-1946, London 1991.

33 *David Matless*, Landscape and Englishness, London 1998.

ral order; Hugh Walpole's popular recreations of a quasi-historical Lake District; Stanley Baldwin's ruminations on what constituted the "real", ineradicable heart of England – each of these was symptomatic of a cultural mood designed to eradicate memories of the Somme and return to an imagined Edwardian idyll[34]. The impact of the Second World War was different. There was no lack of writing expressing a yearning for a rural – aristocratic, hierarchical and "feudal" – past. Evelyn Waugh's *Brideshead Revisited* and *Sword of Honour trilogy* explore precisely this terrain[35]. The predominantly rurally rooted fiction of L.P. Hartley is located in a similar though more worldly cultural milieu, while the extraordinarily popular H.E. Bates and Laurie Lee luxuriated in unabashed rural nostalgia[36]. In general, however, the Second World War generated a desire to forget rather than reimagine, recover or reinvent. The economic disasters of the 1930s, the nadir of urban-industrialism, and failure on the part of successive pre-war governments to frame policies which might have reduced the trauma of mass unemployment, ensured that the dominant mood tended to be forward-looking, disowning rather than clinging to a mythological preindustrial past. Yet this mood did little to weaken the attractions of anti-urban rural preservationism. The election of a Labour government ensured that social policies prepared during the middle years of the war were rapidly implemented[37]. At the same time, planners had now come of age and insisted that the blitz provided an inherently conservative nation with an opportunity to undo what were now conventionally assumed to be the environmental monstrosities of Victorian industrialism[38]. The polluted city would be cleansed by combining architectural modernism with variants on *rus in urbe*. New towns and the rebuilding of centres subjected to heavy bombing would deploy zoning to validate the spatial coexistence of nature and – preferably light – industry. At the same time planners experimented with city-based schemes for suburbs underwritten by the values of hamlet and village[39]. Social and architectural engineering were now thought to be capable of reproducing elements of the rural within modernistic architectural and residential settings[40]. Linked bodies of legisla-

34 *Valentine Cunningham*, British Writers of the Thirties, Oxford 1988, 229-38. The case of Baldwin is intriguingly interrogated in *Philip Williamson*, Stanley Baldwin. Conservative Leadership and National Values, Cambridge 1999. See, also, *J.M. Winter*, Sites of Memory, Sites of Mourning. The Great War in English Cultural History, Cambridge 1995.

35 *R.M. Davis*, Brideshead Revisited. The Past Redeemed, Boston 1990.

36 On Hartley there is *A. Wright*, Foreign Country. The Life of L.P. Hartley, London 1996. See, also, *D.P. Vannatta*, H.E. Bates, Boston 1983 and *Laurie Lee*, Cider with Rosie, London 1967 and As I Walked Out One Midsummer Morning, London 1969.

37 *Paul Addison*, The Road to 1945. British Politics in the Second World War, London 1975.

38 *Alison Ravetz*, Remaking Cities. Contradictions of the Recent Urban Environment, London 1980, 19-24.

39 *P. Richmond*, Marketing Modernisms. The Architecture and Influence of Charles Reilly, Liverpool 2001, 177-98. See, also, *Mark Clapson*, Invincible Green Suburbs, Brave New Towns: Social Change and Urban Dispersal in Interwar Britain, Manchester 1998 and *S. Fielding*, *P. Thompson* and *N. Tiratsoo*, England Arise! The Labour Party and Popular Politics in 1940s Britain, Manchester 1985, 102-34.

40 *Bill Luckin*, History, Community and Utopia: Imagined Futures and the Manchester Plan of 1945: Figures et Strategies D'Elite Technique dans L'Europe après 1945, eds. *P. Bonifazio* and *E.C. Lanza* (Geneva, forthcoming, 2006).

tion would be introduced to protect wildlife, wilderness and places of natural beauty proximate to and far distant from the great city[41].

III

The impact of a dominant discourse is enhanced by failure to engage with a full range of empirical evidence. Commonsensical, natural, *a priori* assumptions interact with linear logical structures to account for no more than a limited spectrum of simplified social and environmental variables to guarantee coherence, credibility and possibility for yet further simplification and popularization. Dominant discourses become even more persuasive if they can be readily absorbed by interest groups able to adapt central tenets to their own professional and social needs. Popularization and simplification also work in favour of effective dissemination throughout the social system as a whole[42]. As with election manifestoes, so also with a dominant discourse, analysis and interpretation are less usefully concerned with evaluating policy solutions to convoluted social and economic problems than the ways in which complexity is rhetorically reduced to the familiar coinage of predictability.

Anglo-centric, pro-rural preservationism in Britain between 1870 and 1945 gained in persuasiveness because it gave little detailed attention to the crucial issue of the material and social formation of an emergent national suburbia. As we have seen, perceived dilemmas associated with the octopus-like growth of the suburbs, particularly in London, generated wide-ranging comment. But throughout our period this debate was abstracted from its concrete economic, social and spatial contexts. Statistical generalization continues to be treacherous but in nearly every major urban centre between the 1860s and the 1940s and beyond suburban outpaced city-wide rates of population growth[43]. Commercialization of the urban core and the construction of central railway termini pushed those who could afford extra outgoings in terms of rent and travel to an inner suburban ring: a significant minority lacking the wherewithal to make such a move became – to use the standard Victorian euphemism – "dishoused" and found themselves forced to endure levels of overcrowding as severe as those experienced in the 1840s[44]. Either that, or they had to find shelter on the unforgiving streets of the great city.

[41] *B.W. Clapp*, An Environmental History of Britain since the Industrial Revolution, London 1994, ch. 6.

[42] *J.B. Thompson*, Ideology and Modern Culture: Critical Social Theory in the Era of Mass Communication, Cambridge 1990.

[43] *Phil Jones*, The Suburban High Flat in the Post-War Reconstruction of Birmingham, 1945-70: Urban History 32 (2005), 308-310.

[44] *D.R. Green*, From Artisans to Paupers: Economic Change and Poverty in London 1780-1870, Aldershot 1995. "Dishousing" is discussed in *J.R. Kellett*, The Impact of the Railways on Victorian Cities, London 1969 and *Gareth Stedman Jones*, Outcast London: A Study in the Relationship between Classes in Victorian Society, Oxford 1971. On the complex balance between low income and rent see *E.H. Hunt*, Regional Wage Variations in Britain, 1850-1914, Oxford 1973.

Every suburban area was different from its near-neighbour or from what might seem to be similar localities in other towns or cities[45]. In broadly schematic terms, outward movement from the inner core gave rise to successive residential rings. These accommodated finely differentiated sub-groups – better-off members of the working-class and the aristocracy of labour, non-manual employees in routine administrative jobs in the central business district, the lower middle-class, the middle class itself and what George Orwell famously termed the lower upper-middle class. At the periphery the affluent upper middle-class inhabited spacious detached villas. These elite members of the population lived in semi-rural splendour in communities retaining many of the characteristics of traditional village life but contained a significantly larger range and proportion of shopkeepers and tradesmen. In London, Hampstead and Highgate were the classic examples[46].

The nationwide shift towards mass suburbanization, which began in London in the earliest years of the nineteenth century, and then replicated itself in Birmingham, Manchester, Leeds and smaller centres, would not have developed in the way it did in the absence of a regionally staggered upward shift in real income from the 1850s onwards[47]. Increased economic stability exerted a profound effect on residential and occupational change within the late Victorian city, and raised reality and expectation in terms of health, environment and amenity[48]. Later in the century, the development of mass transport systems, and particularly of cheap fares, reinforced the suburbanising process[49]. At the same time, those who remained behind at the inner core experienced a relative decline in health and amenity[50]. This was the era of the mass slum as well as of mass suburbanization. Sometimes, as in the capital, the two spheres were organically linked to one another[51].

Both in London and the great provincial centres communities bordering urban regions came gradually to accept that it would be in their own best interests to be formally incorporated into the great city. Thus in the late nineteenth century the villages of Chorlton, Withington and Didsbury allowed themselves to be administratively colonised by Manchester, thereby preparing the way for an even more massive suburban thrust deep into agricultural Cheshire during the inter-war era. In fact, it was during the 1920s and 1930s that residence in a terraced, semi-detached or detached

45 *F.M.L. Thompson*, Introduction: The Rise of Suburbia, ed. F.M.L. Thompson, Leicester 1982, 2.

46 *F.M.L. Thompson*, Hampstead: Building a Borough, 1650-1954, London 1974. See, also, *Jane Austen*, Emma, 1980.

47 This issue has been succinctly summarized by *Hoppen* (note 22), 72-90.

48 New perspectives on the issue of health have been opened up by *Naomi Williams*, Death in its Season; Class, Environment and the Mortality of Infants in Nineteenth Century Sheffield: Social History of Medicine 5 (1992), 71-94.

49 *Thompson* (note 45), 6.

50 *S.R. Szreter* and *G.M. Mooney*, Urbanization, Mortality and the Standard of Living Debate: New Estimates of Expectation of Life at Birth in Nineteenth Century British Cities: Economic History Review lx (1998), 108-110.

51 See *H.J. Dyos* and *D.A. Reeder*, Slums and Suburbs: The Victorian City: Images and Realities, vol 1, eds. H.J. Dyos and Michael Woolf, 359-88 and *Richard Dennis*, Modern London: Cambridge Urban History of Britain, vol 3: 1840-1950, ed. M. Daunton, Cambridge 2000, 117-21.

property, far removed from the environmental *sturm und drang* of the urban core, became a dominant cultural as well as economic and residential norm[52]. At the same time, outsiders, inhabitants of the most deprived and least salubrious areas of cities, whose parents had lacked the wherewithal to escape to a respectable inner suburb, remained trapped in what had come by the 1880s to be known as the shamefully "hidden" slums of London, Manchester, Leeds and Birmingham. They were still there in the 1930s[53]. The only, though frequently economically and culturally problematic, alternative was to move to an over-expensive and socially unwelcoming municipal housing estate[54].

The rush towards suburbia was not, of course, uniform or uniformly rapid. Larger numbers of people in smaller and poorer towns – the communities comprising the Black Country are a good example – which lacked a variegated social and occupational structure and a large and vigorously upwardly mobile professional class, remained clustered closer to the urban core than in large cities[55]. But every centre of any size boasted its own, distinctive suburban space[56]. The process of differentiation accelerated during the interwar period and even more so in the years after 1945.

How, then, did the emergence of a large-scale national suburbia fail significantly to modify the emphases and tonalities of the town-country dichotomy which has been described and interpreted in this essay? First, a venerable urban-rural duality was reinforced by political developments in the early and mid-nineteenth century. Thus in terms of rhetoric, the debate over factory legislation, the crisis surrounding the great Reform Act of 1832 and extended inter- and intra-party conflict over free trade and the Corn Laws in the 1830s and 1840s, were invariably though superficially presented as epochal battles between the landed and the manufacturing classes. This greatly reinforced a cultural tendency to think and generalize within the context of exclusively rural or urban spheres. Second, and as a matter of fact rather than perception or political ideology, the growth of new towns between the 1780s and the 1840s – localities classically named "*shock cities*" by Asa Briggs – had indeed been unprec-

52 The narrative of Mancunian suburban absorption is related in *A. Redford* assisted by *I.S. Russell*, The History of Local Government in Manchester, vol 2: Borough and City, London 1940, 322. For the startling statistics of owner-occupation in England and Wales between the Edwardian period and 1961 see *Colin Pooley*, Patterns on the Ground: Urban Form, Residential Structure and the Social Construction of Space, ed. Daunton (note 51), 445.

53 The major contemporary authority on the inter-war slum was the Mancunian and national Liberal politician and businessman, E.D. Simon. See *E.D. Simon*, How to Abolish the Slums, London 1929, The Anti-Slum Campaign, London 1933 and The Rebuilding of Britain: A Twenty Year Plan, London 1942. See, in addition, *Robert Roberts*, The Classic Slum: Salford Life in the First Quarter of the Century, Manchester 1971.

54 See the exemplary *A. Olechnowicz*, Working-Class Housing in England between the Wars: The Becontree Estate, Oxford 1997.

55 See *James Winter*, Secure from Rash Assault: Sustaining the Victorian Environment, Berkeley 1999, 152-61 and *R.H. Trainor*, Authority and Social Structure in an Industrial Area: A Study of Three Black Country Towns, Oxford 1981.

56 The seminal study of a specific area remains *H.J. Dyos*, Victorian Suburb: A Study of the Growth of Camberwell, Leicester 1961.

edentedly rapid and disorienting[57]. As a consequence, paternalistic conservatives focused almost exclusively on the midlands and north, rather than pockets of abject poverty in much longer established centres or, crucially, on *spatial and occupational differences* between the shock cities themselves. This, also, tended subtly to exclude the suburban variable.

Third, the new industrial cities, which began to move towards a degree of political and administrative maturity in the mid-1830s, felt impelled stridently to assert their emergent identities over and against those of the landed elite – the so-called "feudal" interest. The new and dynamic ideology of municipal government, predominantly articulated by civic leaders preoccupied by the task of reforming specifically inner city infrastructures, proved central to this discourse. As a consequence, the "middle" -the by now rapidly expanding suburban sector – was once again tacitly excluded. Fourth, late nineteenth century and Edwardian metropolitan intellectuals invariably depicted the suburbs as irremediably philistine[58]. This critique combined with incessant attacks on the visual drabness of the new and rapidly expanding sector. The "typical" suburb was said to lack a centre or "heart", display an alienating architectural uniformity, and communicate neither the peace and quiet of the village nor – a paradox, this, given the ubiquity of pro-agrarian ideology – the exhilarating bustle of a crowded metropolitan or urban-industrial street[59].

Finally, and astonishingly, only a minority of nineteenth century commentators drew attention to the fact that, from the very outset, some, though by no means all, suburbs adopted and mimicked the values and life-styles *of* the countryside[60]. Later, of course, it would become a truism that preference for terrace or semi-detached rather than flat or apartment, dedication to garden or allotment, and, as time went on, commitment to bowls, cricket, tennis and Sunday park football, demonstrated enthusiasm not for a rural idyll *sang pure* but for a very particular kind of attachment to the country in the city.

In reality, then, late nineteenth and earlier twentieth century suburbs constituted a bridge between the at least partly imagined polarity of country and city. But the imbedded resilience of a social and cultural discourse reaching back to the late medieval period, and rooted in intense distaste for the immorality, luxury and moral pollution of the capital, militated against a radical modification of the town-country dyad, a di-

57 *Asa Briggs*, Victorian Cities, London 1963.

58 *Carey* (note 29), ch. 3.

59 *James Winter*, London's Teeming Streets, London 1993, provides a compelling overview of the hustle and bustle of the nineteenth century capital. However, a group of privileged upper middle-class inhabitants of the metropolis actively campaigned against the hubbub of the street. See *John M. Picker*, The Sound-Proof Study: Urban Professionals, Work Space and Urban Noise: Victorian Studies 42 (1999-2000), 427-53.

60 From the early years of the nineteenth century onwards well-to-do suburbanites were well provided with advice literature on how to import elements of nature into the city. The pioneer was John Claudius Loudon, editor of the *Gardener's Magazine*, and author of The Suburban Gardener and Villa Companion, London 1838 and The Suburban Horticulturalist, London 1842. See *Melanie Louise Simo*, Loudon and Landscape: From Country Seat to Metropolis 1783-1943, New Haven 1988.

chotomy which, as we have seen, profoundly shaped late nineteenth and twentieth century rural preservationism. This may seem an unremarkable conclusion. But only when the suburban variable has been more convincingly theorized and historicized within this context, will it be possible to move towards an adequate account of an abiding cultural duality in modern and contemporary England and Britain[61].

[61] Comparisons between major English and Scottish urban centres will be central to this task. For Scotland see *Richard Rodger*, Edinburgh and the Transformation of the Nineteenth Century City: Land, Property and Trust, New York and Oxford 2000.

Dieter Schott

Industrialisierung und städtische Umwelt in Deutschland

Fassungslos beobachtete die Welt in den Wochen vor der Tagung in Coburg September 2005, wie New Orleans, eine traditionsreiche und einst hochattraktive Großstadt der USA, durch das Wüten des Wirbelsturms Katrina und die folgenden Deichbrüche am Mississippi in Wassermassen, Schlamm und Dreck versank, wie die städtischen und staatlichen Behörden angesichts des Ausmaßes der Naturkatastrophe offensichtlich zunächst nicht in der Lage waren, rasche Nothilfe zu leisten. Die Erfahrung von New Orleans, und darin liegt die Relevanz für das Thema ‚Industrialisierung und Umwelt', zeigt geradezu paradigmatisch das totale Versagen eines hochgradig integrierten Komplexes technischer Ver- und Entsorgungssysteme, die in ihrem Zusammenwirken die Existenz einer modernen Großstadt an einem hochgradig prekären Standort von New Orleans überhaupt erst ermöglichten[1]. In einem Becken unter dem Meeresspiegel gelegen, von mehreren Seiten von Wasser umgeben, konnte New Orleans nur dank eines Systems von Dämmen und Pumpen existieren, die die Wasserhaltung kontrollierten. Weil Wirbelsturm Katrina nun großflächig an der amerikanischen Golfküste die Stromversorgung sowie auch praktisch alle anderen Ver- und Entsorgungssysteme lahm legte, konnten weder die Dammbrüche repariert noch die Pumpen betrieben werden. Auch unter normalen Umständen hängt das Funktionieren einer komplexen arbeitsteiligen Großstadt, insbesondere in subtropischen Breiten, sehr wesentlich von der Stromversorgung ab, angefangen von Kühlschränken, Klimaanlagen, Alarmanlagen bis hin zu computergesteuerten Registrierkassen und Zahlungssystemen. New Orleans erlebte also gewissermaßen eine kollektive zivilisatorische Regression auf eine Stufe vor Einführung der Wasser-, Gas- und Elektrizitätsversorgung. Dies geschah allerdings, ohne dass die Mitte/Ende des 19. Jahrhunderts gegebenen und durchaus funktionsfähigen Ver- und Entsorgungssysteme vorhanden waren, ohne dass das dazu gehörende Wissen und Knowhow noch gesellschaftlich verankert war.

[1] Zur Umweltgeschichte von New Orleans und der spezifischen Lage der Stadt vgl. Transforming New Orleans and its Environs, ed. *Craig Colten*, Pittsburgh 2000.

1. Wie wirkt sich Industrialisierung auf die städtische Umwelt aus?

Das Thema ,Industrialisierung und städtische Umwelt' wirft Fragen insbesondere auf zwei Ebenen auf: Einmal geht es darum, welche Umwelteffekte Industrialisierung für Städte und deren Umwelt ausgelöst hat. Der zweite Fragenkomplex bezieht sich darauf, in welcher Weise Industrialisierung auch ein technologisches und konzeptionelles Angebot zur Bewältigung städtischer Problemlagen erzeugt hat. Dieser Beitrag setzt sich schwerpunktmäßig mit dem zweiten Komplex auseinander, aber vorab noch ein paar Bemerkungen zu den Umwelteffekten der Industrialisierung.

Der offensichtlichste und wohl auch in der Umweltgeschichte der letzten Jahrzehnte am besten aufgearbeitete Aspekt ist die Umweltverschmutzung durch die Rauchemissionen der Fabriken und Feuerungsstätten, die Emittierung von flüssigen und festen Reststoffen industrieller Produktion, die Gewässer verunreinigten, Böden vergifteten. Ausmaß und Intensität dieser Umweltverschmutzung hing direkt von der Industriedichte bzw. auch von der spezifischen Art der industriellen Produktion ab. Schwerpunktregionen waren daher die industriellen Konzentrationen wie insbesondere das Ruhrgebiet, aber auch industriell geprägte Großstädte wie Berlin. Zu dieser Problematik liegt eine größere Anzahl von empirischen Studien vor[2].

Indirekter, aber vielleicht für die städtische Umwelt insgesamt noch folgenreicher war die mit der Industrialisierung einhergehende Transport-Revolution, d.h. die Maschinisierung des Güter- und Personentransports zu Lande und zu Wasser. Eisenbahn und Dampfschiff erweiterten nicht nur die Mobilitätshorizonte der Menschen exponentiell, sie steigerten auch die Entfernungen, über die Massengüter wie Kohle, Holz, Getreide oder Baustoffe transportiert und noch zu konkurrenzfähigen Preisen am Zielort verkauft werden konnten, innerhalb weniger Jahre um das Mehrfache. Dadurch wurde die Kohle, die vor der Mitte des 19. Jahrhunderts nur in relativ begrenzten Räumen in der Nähe der Kohlegruben als Brennstoff vorherrschte, innerhalb weniger Jahrzehnte in großen Teilen Deutschlands zum universalen Brennstoff nicht nur für Industrie, sondern auch für den öffentlichen und privaten Hausbrand. Der Übergang von Holz als energetischem Grundstoff der vorindustriellen Periode zu Kohle markiert nach Rolf Peter Sieferle auch eine Transformation des Energiesystems von nachwachsenden zu fossilen Rohstoffen, ein im Hinblick auf aktuelle De-

[2] *Arne Andersen*, Historische Technikfolgenabschätzung am Beispiel des Metallhüttenwesens und der Chemieindustrie 1850-1933, (Zeitschrift für Unternehmensgeschichte: Beiheft 90) Stuttgart 1996; *Franz-Josef Brüggemeier*, Das unendliche Meer der Lüfte: Luftverschmutzung, Industrialisierung und Risikodebatten im 19. Jahrhundert, Essen 1996; *Franz-Josef Brüggemeier* und *Thomas Rommelspacher*, Blauer Himmel über der Ruhr. Geschichte der Umwelt im Ruhrgebiet 1840-1990, Essen 1992; *Jürgen Büschenfeld*, Flüsse und Kloaken: Umweltfragen im Zeitalter der Industrialisierung (1870-1918), (Industrielle Welt 59) Stuttgart 1997; *Ulrike Gilhaus*, „Schmerzenskinder der Industrie". Umweltverschmutzung, Umweltpolitik und sozialer Protest im Industriezeitalter in Westfalen 1845-1914, (Forschungen zur Regionalgeschichte 12) Paderborn 1995; *Frank Uekötter*, Von der Rauchplage zur ökologischen Revolution. Eine Geschichte der Luftverschmutzung in Deutschland und den USA, 1880-1970, (Veröffentlichungen des Instituts für Soziale Bewegungen: A, 26) Essen 2003.

batten über Klimawandel, CO_2-Ausstoß und Nachhaltigkeit fundamentaler Vorgang[3].

Für die städtische Topographie brachte die Transportrevolution ebenfalls starke Umwälzungen. Bahnhöfe und Bahnanlagen wurden meist in einiger Entfernung von den traditionellen Stadtzentren errichtet, sie konstituierten neue Wachstumspole; Bahnhofsstrassen, die die alten Stadtkerne mit dem Bahnhof verbanden, wurden häufig zu ersten Adressen. Der Bahnhof wurde das neue „Stadttor", hier kamen die Reisenden an, von hier aus führte der unvermeidliche Baedeker bildungsbeflissene Reisende durch die Sehenswürdigkeiten einer Stadt[4].

Zugleich bildeten sich um die Bahnhöfe und Bahnanlagen wegen der kostengünstigen Rohstoffversorgung und des leichten Zugangs zu auswärtigen Märkten häufig Industrieviertel, die sich auf die städtische Umwelt im Hinblick auf die Wohn- und Standortqualität benachbarter, von den Hauptwinden bestrichener Viertel auswirkten. In stadträumlicher Hinsicht wirkte die Eisenbahn ambivalent. Einerseits fungierte sie als „Stadtöffner", beschleunigte und intensivierte den Prozess der Entfestigung der Stadt, der Öffnung zum Umland. Andererseits bildeten Bahngleise und Bahndämme in dem Maße, wie sich die Bahnverbindungen in die verschiedensten Richtungen erstreckten, schon bald wieder neue Hemmnisse und Strukturierungsfaktoren für das Stadtwachstum[5]. Wohnviertel mit schlechtem Hausbestand und starker Umweltbelastung durch angrenzende Industrie lagen – wie man in England zu sagen pflegte – „on the wrong side of the tracks". Bahnanlagen beanspruchten große Mengen städtischen Bodens. Die Projekte der Deutschen Bahn mit „Bahn 21" in Stuttgart und Frankfurt, die großen Reserven und untergenutzten Areale kernstädtischen Bodens zur Entwicklung der Innenstädte zu mobilisieren, beruhen auf dieser Struktur[6].

[3] Vgl. *Friedrich-Wilhelm Henning*, Die Industrialisierung in Deutschland, Paderborn [3]1976; *Lothar Gall*, Eisenbahn in Deutschland. Von den Anfängen bis zum Ersten Weltkrieg, in: Die Eisenbahn in Deutschland: Von den Anfängen bis zur Gegenwart, hg. v. Lothar Gall und Ralf Roth, München 1999, 13-70; *Rolf-Peter Sieferle*, Der unterirdische Wald. Energiekrise und industrielle Revolution, München 1982; *ders.*, Nachhaltigkeit in universalhistorischer Perspektive, in: Umweltgeschichte. Themen und Perspektiven, hg. v. Wolfram Siemann, München 2003, 39-60; *Joachim Radkau*, Natur und Macht. Eine Weltgeschichte der Umwelt, München 2000.

[4] Aus kulturgeschichtlicher Perspektive immer noch anregend *Wolfgang Schivelbusch*, Geschichte der Eisenbahnreise. Zur Industrialisierung von Raum und Zeit, Frankfurt/M. 1977, bes. 152-157; zum Verhältnis Bahn und Stadt in Deutschland jetzt *Ralf Roth*, Interactions between railways and cities in nineteenth-century Germany: some case studies, in: The City and the Railway in Europe, hg. v. Ralf Roth und Marie-Noëlle Polino, (Historical Urban Studies) Aldershot 2003, 3-27.

[5] Zum Problem der Bahnhofsverlegungen Ende des 19./Anfang des 20. Jahrhunderts vgl. *Wolfgang R. Krabbe*, Frühindustrielle „Altlasten" in der Blütezeit der Urbanisierung: Städtische Eisenbahn- und Bahnhofsprobleme um 1900, in: Stadt und Verkehr im Industriezeitalter, hg. v. Horst Matzerath, (Stadtforschung, A 41) Köln/Wien 1996, 109-129. Am Beispiel Darmstadt habe ich herausgearbeitet, wie eine solche Bahnhofsverlegung zum Knotenpunkt einer Vielzahl infrastruktureller Veränderungen werden konnte: *Dieter Schott*, Die Vernetzung der Stadt. Kommunale Energiepolitik, öffentlicher Nahverkehr und die Produktion" der modernen Stadt. Darmstadt, Mainz, Mannheim 1880-1918, Darmstadt 1999, 265-284.

[6] Vgl. Stuttgart 21: Entwürfe für die neue Stadt, hg. v. der Landeshauptstadt Stuttgart, Stuttgart 1996.

Die zweite Ebene des Komplexes Industrialisierung – städtische Umwelt bezieht sich stärker auf den Angebotscharakter der Industrialisierung, darauf, wie im Zuge der Industrialisierung eine breite Palette von Technologien, von Lösungsmitteln generiert wurde, mit deren Hilfe Städte, ihre Ingenieure und Techniker versuchten, die z.T. selbst industrialisierungsbedingten Probleme der Städte in den Griff zu bekommen. Diese Ebene soll im Folgenden im Mittelpunkt stehen, insbesondere, weil die Problemlösungsstrategien der Industrialisierungsepoche vielfach Strukturen geschaffen haben, die uns heute einerseits als erhebliche Kostenfaktoren in der Bauerhaltung, andererseits als restriktive Rahmenbedingungen beim Versuch einer nachhaltigen Umorientierung der Städte gegenüber stehen[7]. Zeitlich wird der Schwerpunkt auf der formativen Epoche 1850-1914 liegen.

2. Die Umweltkrise als Hygienekrise

Mitte des 19. Jahrhunderts kann man für viele europäische und auch deutsche Städte zweifellos von einer Krise der städtischen Umwelt sprechen, allerdings primär in dem Sinne, dass die städtische gebaute Umwelt der Straßen, Plätze und Wohnhäuser als krankheitsfördernd, als Brutstätten für Seuchen gesehen wurden. Städte hatten traditionell hohe Sterblichkeitsraten und über lange Zeitperioden seit dem Mittelalter konnten Städte nur durch den Zustrom vom Land ihren Bevölkerungsstand halten oder sogar ausbauen[8]. Mitte des 19. Jahrhunderts spitzte sich die Situation angesichts der neuartigen Epidemie Cholera sowie der Bevölkerungsverdichtung in den Städten deutlich zu. Für die Wahrnehmung dieser Krise der städtischen Umwelt wie auch für die Ausarbeitung von Lösungsvorschlägen spielte nun die britische Public-Health-Bewegung, insbesondere ihr Wortführer Edwin Chadwick, auch für Deutschland eine maßgebliche Rolle. Für Chadwick, im Hauptberuf Sekretär der Poor Law Commission, resultierte Armut aus Krankheit, die wiederum als Folge mangelnder Hygiene gesehen wurde. Organischer Unrat aus Essensabfällen, tierischen und menschlichen Fäkalien im öffentlichen wie privaten städtischen Raum erzeugte Faulgerüche, so genannte „Miasmen“, die nicht nur als lästig, sondern als direkte Auslöser von Krankheiten wie Ruhr, Typhus und Cholera galten[9].

Chadwick forderte in seinem berühmten „Sanitary Report“ von 1842 eine durchgehende „Hygienisierung“ der Stadt, einerseits durch die Versorgung der Privathäu-

[7] Vgl. dazu die Beiträge in: Resources of the City, hg. v. *Dieter Schott*, *Bill Luckin* und *Geneviève Massard-Guilbaud*, (Historical Urban Studies) Aldershot 2005, insbesondere die Einleitung: Dieter Schott, Resources of the City: Towards A European Urban Environmental History, 1-27. Vgl. auch zum Reparaturbedarf städtischer Infrastruktur: Informationen zur Raumentwicklung, Heft 5/2006, „Zukunft städtischer Infrastruktur“.

[8] Zur Problematik hoher Mortalität in Städten vgl. *Heinz Schilling*, Die Stadt in der frühen Neuzeit, (Enzyklopädie deutscher Geschichte 24) München 1993, bes. S. 2-17.

[9] *Christopher Hamlin*, Public Health and Social Justice in the Age of Chadwick. Britain, 1800-1854, Cambridge 1998; zur Cholera *Michael Dorrmann*, ‚Das asiatische Ungeheuer‘. Die Cholera im 19. Jahrhundert, in: Das große Sterben. Seuchen machen Geschichte, hg. v. Hans Wilderotter, Dresden 1995, 204-251.

ser mit Wasserleitungen anstelle der häufig verseuchten und verschmutzten Grundwasserbrunnen, andererseits durch die Entsorgung der Fäkalien mittels einer Kanalisation; als Schnittstelle der beiden Systeme diente das seit Beginn des 19. Jahrhunderts rasch populär gewordene Wasser-Closet[10]. Diese Technisierung von Wasserversorgung und Entsorgung der Abwässer griff tief in vorhandene stoffliche Austauschprozesse zwischen Stadt und Umland ein: In der vorindustriellen Stadt waren menschliche und tierische Fäkalien und sonstiger Unrat in der Regel in Abtrittsgruben gesammelt und in regelmäßigen Abständen in das agrarische Umland der Stadt gebracht worden, wo es als Dünger für die Intensivlandwirtschaft Verwendung fand. Weil Mitte des 19. Jahrhunderts weder Guano noch industriell produzierter Kunstdünger zur Verfügung stand, wog der Einwand von Justus von Liebig schwer, die Ableitung von Fäkalien in die Kanalisation stelle einen Raubbau dar. Chadwick stellte dem sein Kreislaufmodell entgegen. Das geplante Netzwerk von Ver- und Entsorgungsleitungen vermittle zwischen Stadt und Land. Wasser diente im Organismus Stadt – in Analogie zum Blut – als Transportmittel. Es sollte die Reststoffe des individuellen und kollektiven Stoffwechsels ohne Gesundheitsrisiken aus dem städtischen Raum entfernen und die im Abwasser enthaltenen organischen Nährstoffe, die der Anbau von Kulturpflanzen zuvor dem Boden entzogen hatte, durch Verrieselung der Landwirtschaft zurückgeben[11].

In Deutschland setzte der große Durchbruch der Hygienebewegung erst im letzten Drittel des 19. Jahrhunderts ein. Eine frühe Ausnahme bildete das Hamburger Wasserversorgungs- und Abwassersystem, das nach dem Großen Brand in Hamburg von 1842 errichtet wurde, begünstigt durch massiven Technologietransfer aus England in Gestalt des Bauingenieurs William Lindley. Das Motiv, hier solche umfangreichen Investitionen zu tätigen, war neben der Hygiene insbesondere auch die Verbesserung des Brandschutzes[12]. In Zusammenhang mit dem starken Wirtschaftswachstum des neu gegründeten Reiches setzten an vielen Orten nach 1870 ernsthafte Bestrebungen ein, die hygienische Situation zu verbessern[13].

[10] *Edwin Chadwick*, Report on the Sanitary Condition of the Labouring Population of Great Britain, hg. v. Michael Pinn, Edinburgh 1968 (urspr. London 1842). Eines der ersten Beispiele, wo die ‚Philosophie' Chadwicks konsequent umgesetzt wurde, ist Croydon in der Nähe Londons, vgl. *Nicholas Goddard*, ‚Sanitate Crescamus': Water Supply, Sewage Disposal and Environmental Values in a Victorian Suburb, in: Resources (wie Anm. 7), 132-148.

[11] *Hamlin* (wie Anm. 9), bes. 237-240. Vgl. zur Debatte über den Wert von Fäkalien *Susanne Hauser*, „Reinlichkeit, Ordnung und Schönheit" – Zur Diskussion über die Kanalisierung im 19. Jahrhundert, in: Die Alte Stadt 23 (1992), 292-312.

[12] Zum Bau der Kanalisation in Hamburg *Volker Roscher*, Die Einführung der Kanalisation in Hamburg und der Neubau der Stadt 1842/43, in: Die Alte Stadt 24/1993, 229-241; *Richard Evans*, Tod in Hamburg. Stadt, Gesellschaft und Politik in den Cholerajahren 1830-1910, Reinbek bei Hamburg 1990.

[13] *John v. Simson*, Kanalisation und Städtehygiene im 19. Jahrhundert, Düsseldorf 1983; *Peter Münch*, Stadthygiene im 19. und 20. Jahrhundert. Die Wasserversorgung, Abwasser- und Abfallbeseitigung unter besonderer Berücksichtigung Münchens, (Schriftenreihe bei der Historischen Kommission der Bayerischen Akademie der Wissenschaften, 49) Göttingen 1993; *Thomas Bauer*, Im Bauch der Stadt: Kanalisation und Hygiene in Frankfurt am Main. 16.-19. Jahrhundert, (Studien zur Frankfurter Geschichte 41) Frankfurt am Main 1998.

Die Vorkämpfer der Verbesserung öffentlicher Hygiene organisierten sich im *„Deutschen Verein für öffentliche Gesundheitspflege“*, darunter prominente Ärzte wie Pettenkofer, Virchow und Varrentrapp, Ingenieure wie James Hobrecht, der ab Ende der 1860er Jahre die Berliner Kanalisation entwarf, oder Reinhart Baumeister, der mit seinem Handbuch von 1876 eine erste wissenschaftliche Anleitung für das Planen und Entwerfen von Stadterweiterungen vorlegte. Es gelang dieser gut organisierten Pressure Group, das Konzept einer systemischen, zentralisierten Wasserver- und Entsorgung bis gegen Ende des 19. Jahrhunderts in den meisten deutschen Städten in die Tat umzusetzen[14]. Ein nicht unerhebliches Motiv für den Bau von Kanalisationen war dabei neben der Verbesserung der städtischen Gesundheitsverhältnisse auch die Gewinnung neuen Baulands durch Trockenlegung durchfeuchteter Stadtteile[15].

Über die Assanierung der Städte hinaus wurde „Hygiene“ im letzten Viertel des 19. Jahrhunderts zu einem Zentralbegriff des öffentlichen Reformdiskurses; im Hinblick auf ihre hygienischen Qualitäten wurden die verschiedensten Dimensionen des gesellschaftlichen und wirtschaftlichen Lebens beleuchtet[16]. Noch heute sind die zahlreichen Bände des Handbuchs der Hygiene, herausgeben von Theodor Weyl in den 1890er und 1900er Jahren zentrale Quellen für Umwelthistoriker, die sich mit Problemlagen, Wahrnehmungsweisen und Lösungsstrategien für Umweltprobleme der Jahrhundertwende befassen[17].

Was waren nun in umwelthistorischer Perspektive die Folgen der Hygienisierung? Im Hinblick auf die Gesundheit der Stadtbewohner war der Prozess insgesamt sehr erfolgreich. Ende des 19. Jahrhunderts waren Morbiditäts- und Mortalitätsraten deutlich reduziert, wobei allerdings in der Forschung umstritten ist, welcher Anteil hier der Sanitärhygiene und welcher der verbesserten Ernährungslage zuzuschreiben ist. Nach 1900 lag sogar die Mortalität in der Stadt niedriger als auf dem Land[18].

14 *Juan Rodriguez-Lores*, Stadthygiene und Städtebau: Zur Dialektik von Ordnung und Unordnung in den Auseinandersetzungen des Deutschen Vereins für Öffentliche Gesundheitspflege 1868-1901, in: Städtebaureform 1865-1900: Von Licht, Luft und Ordnung in der Stadt der Gründerzeit, hg. v. Juan Rodriguez-Lores und Gerhard Fehl, (Stadt – Planung – Geschichte 5) Hamburg 1985, 19-58; *Gerd Albers*, Zur Entwicklung der Stadtplanung in Europa, (Bauwelt Fundamente 117) Braunschweig/Wiesbaden 1997, 35 f.; *Anthony Sutcliffe*, Towards the Planned City: Germany, Britain, the United States and France, 1780-1914, (Comparative Studies in Social and Economic History 3) Oxford 1981, 9-27; *Brian Ladd*, Urban planning and civic order in Germany, 1860-1914, (Harvard historical studies) Cambridge, Mass. 1990.

15 *Ulrich Koppitz*, Constructing Urban Infrastructure for Multiple Resource Management. Sewerage Systems in the Industrialization of the Rhineland, Germany, in: Resources (wie Anm. 7), 168-184.

16 *Alfons Labisch*, Homo hygienicus. Gesundheit und Medizin in der Neuzeit, Frankfurt/Main, New York 1992; auch Joachim Radkau sieht in der Hygienebewegung des späten 19. Jahrhunderts eine wirkungsmächtige und durchsetzungskräftige gesellschaftliche Bewegung, vgl. *Radkau*, Natur (wie Anm. 3), 280 f.

17 *Theodor Weyl* (Hg.), Handbuch der Hygiene, 10 Bde, 1893-1904.

18 Vgl. *Jörg Vögele*, Sozialgeschichte städtischer Gesundheitsverhältnisse während der Urbanisierung, (Schriften zur Wirtschafts- und Sozialgeschichte 69) Berlin 2001; *ders.*, Sanitäre Reformen und der Wandel der Sterblichkeitsverhältnisse in deutschen Städten, 1870-1913, in: VSWG 80 (1993), 345-365.

Im Hinblick auf die natürliche Umwelt der Stadt bedeutete die Technisierung letztlich eine Externalisierung der Umweltbelastung, denn Chadwicks ursprüngliches Kreislaufmodell wurde in der Praxis aus finanziellen Gründen oder mangels geeigneten Geländes häufig umgangen. Weil der Bau von Rieselfeldern sehr aufwändig war und auch nicht überall geeignete sandige Böden zur Verfügung standen, wurden Abwässer meist mehr oder weniger ungeklärt in die Flüsse eingeleitet. Diese Immission von Abwasser in die Flüsse produzierte erhebliche politische Konflikte: 1877 verbot Preußen die Einleitung ungeklärter Abwässer in die Flüsse. Daraufhin profilierte sich der *‚Deutsche Verein für öffentliche Gesundheitspflege'* als Vorreiter der Kanalisationslobby, forderte die Rücknahme des Einleitungsverbots, den Bau von Kanalisationen und die Festlegung von Grenzwerten. Reinhart Baumeister sprach vom natürlichen Recht der Städte, die Flüsse als Vorfluter zu benutzen[19].

Eine Minderheit des Vereins spaltete sich daraufhin ab und gründete den *„Internationalen Verein gegen Verunreinigung der Flüsse, des Bodens und der Luft"*. Trotz des universalen Anspruchs im Namen konzentrierte sich der stark auch mit landwirtschaftlichen Interessen assoziierte Verein auf den Kampf gegen die Einleitung von mit Fäkalien belasteten Abwässern in die Flüsse. Dies geschah nicht primär aus umweltschützerischer Motivation, sondern weil die Einleitung als eine Vergeudung wertvollen Düngers, als Nachteil für die Landwirtschaft gesehen wurde[20].

Letzten Endes setzten sich die Kanalisationsbefürworter durch, immer mehr Städte bauten Kanalisationen, leiteten die Abwässer in die Flüsse und der Zustand der Flüsse verschlechterte sich gegen Ende des 19. Jahrhunderts erheblich, was neben den Haushaltsabwässern auch an der Immission industrieller Schadstoffe und an den Abwässern der Kaliindustrie lag[21].

Begründet wurde die Nutzung der Flüsse als Deponie für flüssigen Abfall einerseits mit der Theorie der Selbstreinigungskraft der Flüsse: Es war seit langem beobachtet worden, dass nach Einleitung städtischer Abfallstoffe, etwa aus Gerbereien, Abdeckereien etc. die Flüsse nach einer gewissen Strecke wieder klar waren und das Flusswasser auch keine geschmackliche und geruchliche Beeinträchtigung mehr aufwies. Flüsse konnten sich also tatsächlich selbst reinigen, wobei die mikrobiologischen Mechanismen dieses Vorgangs den Zeitgenossen vor den Entdeckungen Kochs noch unklar waren, vielmehr nahm man eine Art mechanischer Reinigung durch die Unebenheiten des Flussbettes an[22].

Der zweite Grund für das bedenkenlose Verschmutzen der Flüsse lag in der lange Zeit vorherrschenden miasmatischen Krankheitstheorie, die eine Übertragung von Krankheiten durch das Wasser, also etwa durch verseuchtes Trinkwasser, ausschloss. Erst die Choleraepidemie von Hamburg 1892 brachte den endgültigen Beweis, dass

19 Vgl. dazu insbesondere *Jürgen Büschenfeld*, Flüsse (wie Anm. 2), 51-99.

20 Vgl. zum ‚Internationalen Verein' *Büschenfeld*, Flüsse (wie Anm. 2), 79-87 und *John von Simson*, Die Flußverunreinigungsfrage im 19. Jahrhundert, in: VSWG 65 (1978), 370-390.

21 *Büschenfeld*, Flüsse (wie Anm. 2).

22 Büschenfeld zeigt, wie die Hygieniker des Deutschen Vereins für öffentliche Gesundheitspflege in den 1870er Jahren diese These von der Selbstreinigungskraft gezielt gegen mögliche Auflagen zur Reinigung der in die Flüsse geleiteten Abwässer entwickelten und propagierten, *ebd.*, 123 ff.

Trinkwasser Krankheiten, in diesem Falle Cholera, übertragen konnte. Fast 8.000 Menschen starben in der Hansestadt durch verseuchtes Trinkwasser, weil Krankheitskeime mit dem Abwasser in die Elbe geschwemmt waren. Von dort gelangten sie wegen der Tidewirkung wieder in das Wasserwerk, das keine ausreichende Filtration besaß. In der unmittelbar angrenzenden preußischen Stadt Altona, wo das Elbwasser durch eine Sandfiltration gereinigt wurde, gab es dagegen praktisch keine Todesfälle[23].

Gegen Ende des 19. Jahrhunderts zeichnete sich dann ein allmählicher Wandel im Umgang mit Flüssen ab. Als die Stadt Mannheim etwa 1895 ihre Abwässer in den Rhein einleiten wollte, ging die rheinabwärts gelegene Stadt Worms, die ihr Trinkwasser aus dem Rhein bezog, massiv dagegen vor. Nach mehrjährigen Auseinandersetzungen wurde Mannheim schließlich gezwungen, eine mechanische Klärung der Abwässer vorzunehmen, die die Größe der Partikel unter die Schwelle visueller Wahrnehmbarkeit senkte. Ziel der Reinigungsbemühungen war also in erster Linie zu verhindern, dass die Bevölkerung den Rhein als Kloake wahrnahm. Der Stadt Worms wurde im Gegenzug empfohlen, ihre Trinkwassergewinnung aus dem Rhein durch andere Technologien zu ersetzen; der Rhein galt offensichtlich zu diesem Zeitpunkt nicht mehr als eine Quelle gesundheitlich unbedenklichen Trinkwassers. England diente in diese Debatten als Schreckbild. Eine eigens berufene Expertenkommission verlangte, die Abwässer zu klären und warnte vor Zuständen, *„wie sie anderswo z.B. in England vielfach vorgekommen sind“*[24].

Hinsichtlich des Inputs, der Gewinnung von Trinkwasser, leitete die Hygienisierung eine technische Überformung natürlicher Wasserkreisläufe in der Stadt ein. Zur Hygienisierung gehörte auch die Herstellung fester und damit leicht zu reinigender Oberflächen für Straßen und Wege[25]. Wegen dieser Flächenversiegelung konnte der natürliche Niederschlag kaum mehr in den Boden versickern und wurde umgehend in die Kanalisation abgeführt. Anstelle des Zugriffs auf lokales Grundwasser in Form der traditionellen Schöpfbrunnen wurde städtisches Trinkwasser von den Wasserwerken meist von nichtkontaminiertem Grund- oder Quellwasser aus dem Umland der Städte bezogen. Mit der Technisierung der Wasserkreisläufe war eine enorme quantitative Steigerung des Wasserdurchsatzes verbunden, nicht zuletzt, weil das System der Schwemmkanalisation einen erheblichen Wasserdurchlauf erfordert, um die Verschlammung der Kanäle zu verhindern. Der Wasserverbrauch pro Kopf stieg im Vergleich zur Zeit der Versorgung aus Schöpfbrunnen auf rund das Zehnfache. Dies bedeutete angesichts des raschen Bevölkerungswachstums der meisten Städte aber auch, dass die lokalen Vorkommen nicht mehr ausreichten, Trinkwasser musste

[23] *Evans*, Tod (wie Anm. 12).

[24] Stadtarchiv Mannheim A 20/23. Kanalisation der Stadt Mannheim. Ergänzendes Sachverständigen-Gutachten, 12.3.1899, 5. *Dieter Schott*, Remodeling „Father Rhine“: the Case of Mannheim 1825-1914, in: Water, Culture and Politics in Germany and the American West, hg. v. Susan Anderson und Bruce Tabb, (German Life and Civilization 36) New York u.a. 2001, 203-225. Zur Flussverunreinigung in England vgl. *Lawrence E. Breeze*, The British Experience with River Pollution, 1865-1876, New York u.a. 1993.

[25] *Gottfried Hösel*, Unser Abfall aller Zeiten. Eine Kulturgeschichte der Städtereinigung, München 1987.

aus immer größeren Entfernungen herangeschafft werden, Talsperren schufen künstliche Seen als Trinkwasserspeicher. Die Beziehungen zwischen Stadt und ihrem Hinterland gestalten sich durch wasserwirtschaftliche Motive neu; an die Stelle der lokalen oder kleinräumigen Versorgung trat in mittelfristiger Perspektive eine großräumige Zuordnung von Wassergewinnungszonen – meist in dünn besiedelten Mittelgebirgsregionen oder im Alpenraum – und Wasserverbrauchszonen der Großstädte und städtischen Agglomerationen. Große Teile der Bundesrepublik sind heute durch diese Wasserwirtschaft dauerhaft in ihrer lokalen Ökologie verändert[26].

3. Stadttechnik als Umweltproblem und Lösung für Umweltprobleme

Industrialisierung wies also für die städtische Umwelt einen Doppelcharakter auf: einerseits wurden durch ihre ökonomischen und demographischen Effekte Migration in die Städte verstärkt und damit die bereits vorindustriell vorhandenen hygienischen Problemlagen verschärft. Andererseits entfaltete Industrialisierung ein breites Spektrum von Technologien für eben diese Problemlagen, in erster Linie Dampfmaschinen, Pumpen sowie Röhren zum Transport flüssiger und gasförmiger Stoffe. Ein ähnlicher Doppelcharakter tritt uns auch bei anderen Elementen industrieller Städtetechnik entgegen.

Gasbeleuchtung verbesserte die Lichtverhältnisse im öffentlichen städtischen Raum sowie in großen Innenräumen signifikant, gasbetriebene Motoren, Gasbadeöfen oder Gasherde erzeugten lokal wesentlich weniger Rauch und Lärm als kohlegefeuerte Dampfmaschinen, Boiler oder Herde. Aber das Gaswerk als chemische Fabrik, das neben Kohlengas eine breite Palette chemischer Nebenprodukte erzeugte, zählte meist zu den schlimmsten Umweltverschmutzern in den Städten. Gaswerksstandorte waren nach Aufgabe der Gaserzeugung meist hochgradig kontaminiert[27]. Städte entwickelten hier häufig ähnliche Problemlösungsmuster: Wenn Gaswerke angesichts des Stadtwachstums zunehmend ihr unmittelbares Umfeld beeinträchtigten, andererseits wegen steigender Nachfrage Betriebsvergrößerungen anstanden, wurden neue Gaswerke am Rand der Städte in verkehrsgünstiger Lage etabliert, wobei bürgerliche Stadtviertel meist verschont blieben[28].

26 *Thomas Kluge*, Wasser und Gesellschaft. Von der hydraulischen Maschine zur nachhaltigen Entwicklung, (Soziologie und Ökologie 3) Opladen 2000, 70-76; *Thomas Kluge* und *Engelbert Schramm*, Wassernöte. Zur Geschichte des Trinkwassers, Köln 1988.

27 Vgl. zur Geschichte der Gaswirtschaft in Europa, auch mit bibliographischen Überblicken: L'Industrie du Gaz en Europe aux XIXe et XXe Siècles. L'innovation entre marchés privés et collectivités publiques, hg. v. *Serge Paquier* und *Jean-Pierre Williot*, (Euroclio 20) Bruxelles u.a. 2005. Zur kulturellen Bedeutung der Gasbeleuchtung *Wolfgang Schivelbusch*, Lichtblicke. Zur Geschichte der künstlichen Helligkeit im 19. Jahrhundert, München/Wien 1983.

28 Für Mannheim und Darmstadt *Schott*, Vernetzung (wie Anm. 5), 256 u. 423-434; *Dieter Schott*, The Formation of an Urban Industrial Policy to counter pollution in German Cities (1890-1914), in: Le Démon moderne. La pollution dans les sociétés urbaines et industrielles d'Europe/The Modern Demon. Pollution in Urban and Industrial European Societies. hg. v. Christoph Bernhardt und Geneviève Massard-Guilbaud, Clermont-Ferrand 2002, 311-332.

Im Umgang mit der Elektrizität, der Ende des 19. Jahrhunderts bedeutsamsten Stadttechnik, spielten Umweltgesichtspunkte dann eine wesentliche Rolle. Elektrische Beleuchtung versprach Abhilfe gegen die Mißstände und Risiken der Gasbeleuchtung, insbesondere deren Feuergefährlichkeit, Luftverschlechterung und Beschädigung von Kleidern und Raumtextilien. Weil die Gleichstromtechnik allerdings eine Stromerzeugung in unmittelbarer Nähe des Stromverbrauchs erfordert, musste die erste Generation von Kraftwerken in den Zentren der Städte platziert werden, erzeugte somit Rauch, Lärm, Erschütterungen und erhebliche Transportbelastungen[29].

Die berühmte Internationale Elektrotechnische Ausstellung von Frankfurt 1891, auf der sich das Drehstromsystem mit einer spektakulären Fernübertragung elektrischer Energie von Lauffen am Neckar nach Frankfurt erstmals der deutschen Öffentlichkeit präsentierte, ging unmittelbar auf die Suche der Stadt Frankfurt nach technischen Alternativen zu einem als unerwünscht gesehenen Kraftwerk im Stadtzentrum zurück[30].

Die Übertragbarkeit elektrischen Stroms über große Entfernungen inspirierte die Zeitgenossen zu großen utopischen Hoffnungen auf eine mögliche Dezentralisierung von Industrie und Arbeiterschaft durch die neuen Potentiale der Elektrotechnik. Der Frankfurter Oberbürgermeister Franz Adickes formulierte die Hoffnungen auf das neue „Zeitalter der Elektrizität" in seinem Grußwort so:

„Wenn es gelingt, einen Theil der Sünde, welche das Zeitalter des Dampfes an der Menschheit verschuldet hat, im Zeitalter der Elektricität wieder gut zu machen – denn die Individualität des Menschen geräth in Gefahr –, wenn wir dem Einzelnen in sein Haus und seine Werkstätte die theilbare Kraft hineinleiten, so wird das eine Errungenschaft sein, wie sie in der Weltgeschichte noch nicht da war"[31].

Diese fast millenaristisch anmutenden Hoffnungen auf eine Lösung der Probleme des Industriekapitalismus durch die Elektrotechnik war kein Einzelfall. Ende des 19. Jahrhunderts kann man von einer weit verbreiteten Kritik an der Industriegesell-

[29] Zu den Nachteilen des Gaslichts vgl. *Schivelbusch*, Lichtblicke (wie Anm. 27), *Hans-Joachim Braun*, Gas oder Elektrizität? Zur Konkurrenz zweier Beleuchtungssysteme, 1880-1914, in: Technikgeschichte 47 (1980), 1-19. Zu den Systemzwängen der unterschiedlichen Technologien *Helmut Lindner*, Strom. Erzeugung, Verteilung und Anwendung der Elektrizität, (Kulturgeschichte der Naturwissenschaften und der Technik) Reinbek b. Hamburg 1985; *Thomas Herzig*, Wirtschaftsgeschichtliche Aspekte der deutschen Elektrizitätsversorgung 1880 bis 1990, in: Die Geschichte der Stromversorgung, hg. v. Wolfram Fischer, Frankfurt/Main 1992, 123-166. Zu Umweltbelastungen am lokalen Beispiel *Schott*, Vernetzung (wie Anm. 5), bes. 193.

[30] Zur Vorgeschichte der Frankfurter Ausstellung vgl. *Jürgen Steen*, Voraussetzungen einer epochalen Ausstellung, in: „Eine neue Zeit ...". Die Internationale Elektrotechnische Ausstellung 1891, hg. v. Historisches Museum Frankfurt am Main, Frankfurt/Main 1991, 11-23.

[31] *Franz Adickes*, (Trinkspruch bei der Eröffnungsfeier der Internationalen Elektrotechnischen Ausstellung, Frankfurt 16.5.1891), in: Offizieller Bericht über die Internationale Elektrotechnische Ausstellung in Frankfurt am Main, Bd. I, hg. v. Ausstellungsvorstand, Frankfurt am Main 1893, 35. Zu sozialen Utopien, die sich mit der Elektrotechnik verbanden, vgl. auch *Dieter Schott*, Das Zeitalter der Elektrizität: Intentionen – Visionen – Realitäten, in: Jahrbuch für Wirtschaftsgeschichte, 2/1999, S. 31-49.

schaft und der industriellen Großstadt im deutschen Bürgertum ausgehen[32]. Diese Kritik machte sich nicht nur an den sozialen Verhältnissen, vor allem am raschen Wachstum einer Klasse besitz- und wurzelloser Proletarier fest, die als grundlegende Bedrohung gesellschaftlicher Stabilität gesehen wurde. Vielmehr prangerte sie auch die gesundheitsgefährdenden Lebensbedingungen in den verrauchten und verrußten Großstädten des späten 19. Jahrhunderts an, die Häßlichkeit der Mietskasernenviertel und der Industrievorstädte. Die Lebensreform-Bewegung läßt sich in ihrer Suche nach naturnäheren Lebensformen als ästhetische Ausdrucksform der Zivilisations- und Großstadtkritik begreifen, als temporäre Flucht „Aus grauer Städte Mauern", wie sie etwa im Rahmen des „Wandervogel" propagiert wurde.

In der ganz überwiegend vom lokalen Bürgertum bestimmten Stadtpolitik der Jahrhundertwende läßt sich dagegen die Formierung eines Bündels pragmatischer Maßnahmen feststellen. Sie sollten den problematischen Aspekten der Industrie begegnen, insbesondere den die Gesundheit der Stadtbewohner bedrohenden Emissionen. Leitidee dieser Politik war die räumliche Trennung von Industrie – als Hauptquelle der Verschmutzung – und Wohnungen. Wenn Wohnungen weit genug von Fabrikschloten entfernt lägen, so diese Vorstellung, würden die negativen Auswirkungen industrieller Emissionen auf die Gesundheit der Einwohner durch Verdünnung ausreichend reduziert. Instrument dieser Leitidee wurde dann wesentlich die Stadtplanung, insbesondere durch die seit den 1890er Jahren sich entfaltende Idee der Zonenplanung. Josef Stuebben hatte in seinem 1890 erschienenen Handbuch „Der Städtebau" die Notwendigkeit einer Differenzierung von Stadtvierteln nach Nutzungsarten hervorgehoben[33]. Franz-Josef Brüggemeier betont, dass nach einem bahnbrechenden Urteil des preußischen Oberverwaltungsgerichts von 1889 viele Stadtverwaltungen Ortsstatute erließen, die Fabriken von bestimmten Teilen der Gemarkung ausschlossen. Zwar scheiterte der radikale Ansatz des Villenvororts Steglitz bei Berlin, Fabriken auf Steglitzer Gemarkung insgesamt zu verbieten, aber Frankfurts Bauzonenplan von 1891 mit der Unterscheidung zwischen industriefreien, gemischten und der Industrie vorbehaltenen Zonen sollte sich als Vorbild durchsetzen[34].

Neben bau- und genehmigungsrechtliche Instrumente traten in den 1890er Jahren zunehmend aber auch die lokale und regionale Verkehrspolitik, die Energiepolitik, die städtische Bodenpolitik sowie erste Ansätze einer Grünflächen- und Parkplanung. Städte begannen zunehmend, Industrieviertel planerisch auszuweisen, meist in der

[32] Vgl. Die Stadt als Moloch? Das Land als Kraftquell? Wahrnehmungen und Wirkungen der Großstädte um 1900, hg. v. *Clemens Zimmermann* und *Jürgen Reulecke*, Basel u.a. 1999; *Rolf Peter Sieferle*, Fortschrittsfeinde? Opposition gegen Technik und Industrie von der Romantik bis zur Gegenwart, München 1984.

[33] *Josef Stuebben*, Der Städtebau. Handbuch der Architektur, Darmstadt 1890.

[34] *Franz-Josef Brüggemeier*, Umweltprobleme und Zonenplanung in Deutschland. Der Aufstieg und die Herrschaft eines Konzepts 1800-1914, in: Environmental Problems in European Cities in the Nineteenth and Twentieth Centuries/Umweltprobleme in europäischen Städten des 19. und 20. Jahrhunderts, hg. v. Christoph Bernhardt, Münster 2001, 143-163, bes., 155-57; *Andreas Weiland*, Die Frankfurter Zonenbauordnung von 1891 – eine fortschrittliche Bauordnung? Versuch einer Entmystifizierung, in: Städtebaureform 1865-1900 (wie Anm. 14), 343-388.

Nähe der Bahnanlagen oder mit speziellen Industriegleisen ausgerüstet. Wo die Topographie dies zuließ, wurde darauf geachtet, die Industrieviertel im Osten der Städte zu platzieren, damit die Emissionen von den vorherrschend westlichen Winden von der Stadt und den Wohngebieten weggeführt würden. Die gesetzgeberische Handhabe dazu war schon in der deutschen Gewerbeordnung von 1869/1871 enthalten, deren § 23, Abs. 3 die Staaten zum Erlaß entsprechender Landesgesetze ermächtigte[35]. Stadtverwaltungen prämierten nun Industrieansiedelung in diesen Vierteln und unterstützten den Umzug von bestehenden Betrieben durch günstigere Bodenpreise, attraktive Energieangebote und infrastrukturelle Erschließungsmaßnahmen.

So bot die Stadt Mannheim ab 1897 in ihrem neu geschaffenen Industriehafen Grundstücke mit direktem Wasser- und Gleisanschluß an. Zur Energieversorgung wurde Strom vom neu erbauten städtischen Kraftwerk zu Vorzugspreisen angeboten, um für die ansiedlungswilligen Betriebe ein eigenes Kraftwerk von vornherein unwirtschaftlich erscheinen zu lassen und damit die Vielzahl von Dampfkesseln zu reduzieren. Und für die Arbeitskräfteversorgung – Wohnungsbau gab es nicht im Industriehafen – wurde eine Straßenbahnlinie eingerichtet, die Arbeiter vor Schichtbeginn und nach Schichtende zu Vorzugstarifen benutzen konnten[36].

Begünstigt wurden diese Bestrebungen der Stadtverwaltungen durch den starken industriellen Wachstumsschub der Periode 1895-1913, der viele Betriebe zu Standortüberlegungen zwang. Waren attraktive Grundstückspreise und niedrige Strom- und Gastarife für Betriebe ein Zuckerbrot[37], so konnte die Genehmigungspraxis der Behörden im Gegenzug als Peitsche wirken. Zwar genossen einmal genehmigte Dampfmaschinen auch im innerstädtischen Bereich Bestandsschutz, aber eine Erweiterung oder ein wesentlicher Umbau waren genehmigungspflichtig, und daher konnten die Behörden durch Verweigerung solcher Genehmigungen, durch Verzögerung oder bestimmte technische Auflagen einen Druck in Richtung auf Verlagerung ausüben[38]. Wie weit solche Interventionspotentiale tatsächlich genutzt wurden, hing stark vom jeweiligen örtlichen Meinungsklima, der Macht industrieller Interessenvertreter in lokalen Selbstverwaltungsgremien und ähnlichen Faktoren ab. Frank Uekötter hat in seiner Geschichte der Luftverschmutzung in Deutschland und den USA große Unterschiede festgestellt, wie die lokalen Behörden jeweils mit der Rauchfrage umgingen: Insbesondere in Hamburg, Dresden und München zeigen sich längerfristige und systematische Anstrengungen seitens städtischer Behörden oder privater Vereine, die allgemeine Rauchbelastung durch technische Beratung, Überwachung von Feuerungsanlagen und Auflagen signifikant zu reduzieren[39]. Bemerkenswert ist,

35 Siehe dazu *Schott*, Formation (wie Anm. 28), 315 und *Dr. Lent*, Örtliche Lage der Fabriken in Städten. Inwieweit hat sich ein Bedürfnis herausgestellt, von der Bestimmung des § 23, Abs. 3 der Deutschen Gewerbeordnung Gebrauch zu machen?, in: Deutsche Vierteljahrsschrift für öffentliche Gesundheitspflege 21 (1889), 42-70.

36 *Schott*, Vernetzung (wie Anm. 5), 352-396.

37 Dies zeigt etwa die Verlagerung der Firma Merck in Darmstadt, die unter wesentlicher Hilfe der Stadt Darmstadt erfolgte, vgl. *Schott*, Formation (wie Anm. 28), bes. 322-23.

38 Ebd., 324-25.

39 *Uekötter*, Rauchplage (wie Anm. 2).

dass es sich bei allen drei Städten nicht um klassische Industriestädte handelt, sondern diese Städte in erheblicher Weise von ihrer Rolle als repräsentative Residenz- und Handelsstädte mit einem stadtstolzen Bürgertum geprägt waren[40].

Obwohl Umweltprobleme der deutschen Städte also durchaus wahrgenommen wurden und insbesondere die Rauchfrage von der Öffentlichkeit, auch im Hinblick auf chronische Atemwegserkrankungen wie Bronchitis und Tuberkulose, als gravierendes Problem gesehen wurde, läßt sich eine wirksame Bekämpfung des Problems an der Wurzel während dieser Periode nicht feststellen. Uekötter erklärt das „Scheitern" der deutschen Rauchbekämpfung im Vergleich zur amerikanischen Anti-Smoke-Bewegung vor dem Ersten Weltkrieg mit der relativen Stärke deutscher Bürokratie, die es paradoxerweise verhinderte, dass sich eine breite gesellschaftliche Reformbewegung des Problems annahm. Im Gegensatz zu konkreten Einzelfällen war die Verwaltung nicht in der Lage, das diffuse Rauchproblem wirksam zu lösen[41].

Städtische ‚Umweltpolitik', auch wenn diese damals natürlich nicht so genannt wurde, zielte im Kern darauf ab, durch räumliche Trennung von Industrie und Wohnen die Belastung in ausgewählten Teilen der Stadt zu reduzieren. Sie nahm damit letztlich die Konzentration von Umweltbelastungen in bestimmten Stadtvierteln oder gar in größeren Regionen wie dem Ruhrgebiet in Kauf. So wie für die Flussverschmutzung ‚Opferstrecken' definiert wurden, Abschnitte des Flusses, die der industriellen Vernutzung überlassen wurden[42], so gab es auch Stadtteile und ganze Bezirke, wo die meist ärmere Bevölkerung keinen wirksamen Schutz genoss. Ihre Lebensbedingungen unterlagen der „ortsüblichen Belastung", ein juristischer Terminus, mit dem sehr häufig Klagen gegen stark emittierende Betriebe abgewehrt wurden. Es zeigt sich so eine sozial einseitige Belastung mit negativen Umwelteffekten in Vierteln mit vorwiegend Unterschichtenbevölkerung[43].

[40] Auch die Studie von Stephen Mosley über Luftverschmutzung in Manchester belegt, dass es nicht die industriellen Pioniere waren, die besonders früh und aufgeschlossen auf Luftverschmutzung reagierten, *Stephen Mosley*, The Chimney of the World: a history of smoke pollution in Victorian and Edwardian Manchester, Cambridge 2001.

[41] *Uekötter*, Rauchplage (wie Anm. 2), 144-47; generell waren es eher Residenz- und insbesondere traditionelle Handelsstädte wie Frankfurt, Leipzig, Köln oder Mannheim, die gegen Ende des 19. Jahrhunderts besonders profilierte und interventionsbereite Leistungsverwaltungen ausbildeten, vgl. zur Leistungsverwaltung auch *Jürgen Reulecke*, Einleitung, in: Die Stadt als Dienstleistungszentrum, hg. v. Jürgen Reulecke, St. Katharinen 1995, 1-17.

[42] Zum Begriff der ‚Opferstrecke' am Beispiel des Rheins *Mark Cioc*, The Rhine. An eco-biography 1815-2000. (Weyerhaeuser Environmental Books) Seattle 2002, bes. Kap. 5 „Sacrificing a River", 109-143.

[43] Die Wirkung des juristischen Begriffs ‚ortsübliche Belastung' betont insbesondere *Brüggemeier/Rommelspacher*, Blauer Himmel (wie Anm. 2). Die Autoren sprechen hier sogar von einer „Industrieschutzzone Ruhrgebiet", d.h. die Industrie genoß im Ruhrgebiet besonderen Schutz vor polizeilichen und behördlichen Auflagen, 47-49.

4. Potential und Grenzen städtischer ‚Umweltpolitik' zu Beginn des 20. Jahrhunderts

Trotz dieser sozialen Schieflage ist nicht zu bestreiten, dass deutsche Großstädte, allerdings mit sehr erheblichen Unterschieden, zu Beginn des 20. Jahrhunderts Umweltprobleme recht intensiv diskutierten. Auf vielen Feldern waren sie um Abhilfe bemüht, etwa mittels städtischer Maschineninspektionen. Allerdings gaben diese Maßnahmen eindeutig der menschlichen Gesundheit und dem Schutz des Eigentums Vorrang; ein Eigenwert des Schutzes der natürlichen Mitwelt wurde nicht generell akzeptiert. Es vollzog sich in dieser Phase auch die Entfaltung städtischer Leistungsverwaltung, das geradezu explosionsartige Anwachsen kommunalen Personals in einer Vielzahl von Feldern der Daseinsfürsorge mit dem generellen Ziel, Lebensbedingungen in der Stadt zu verbessern, Produktionsbedingungen für die in der Stadt ansässigen Betriebe zu sichern und die Bedrohung gesellschaftlicher Stabilität zu verhindern. Insgesamt zeigt die Debatte über Umweltprobleme in dieser Phase kurz vor 1914 eine gesellschaftliche Breite und Intensität, die erst nach 1970 wieder erreicht wurde; nach Franz-Josef Brüggemeier unterbrach hier möglicherweise der Erste Weltkrieg eine Entwicklung zu effektiverem umweltpolitischen Handeln. So konnte etwa die temporäre Erholung der Umwelt während des passiven Widerstands anläßlich der Ruhrgebietsbesetzung 1923 nur als Bestandteil der gesellschaftlichen Katastrophe wahrgenommen werden; eine allgemeine Erleichterung ging durch das Ruhrgebiet, als ab Ende 1923 die Schlote wieder qualmten, die Kumpels wieder einfahren konnten, aber die Äpfel in den Zechensiedlungen wieder nur kümmerlich wuchsen[44].

[44] *Franz-Josef Brüggemeier*, Tschernobyl, 26. April 1986. Die ökologische Herausforderung, (20 Tage im 20. Jahrhundert) München 1998, 129-30. Die bei *Brüggemeier/Rommelspacher* (wie Anm. 2) abgedruckte Quelle von H. Bergerhoff, Untersuchungen über die Berg- und Rauchschädenfrage mit besonderer Berücksichtigung des Ruhrbezirks, Bonn 1928, belegt eindrucksvoll die deutliche Erholung der Vegetation im raucharmen Sommer 1923, 171-175.

Ian Simmons

Environmental Protection in the UK during the Era of Fossil Fuels

The history of selective environmental protection in the UK is as long as in many other western and oriental cultures. As with many of them, the preservation of wild animals for hunting by a ruling class is confirmed in writing and graphically during the era of pre-industrial agriculture. There is palaeoecological evidence for the selective husbandry of wild species of trees by coppicing since a variety of poles and withies had many uses, and the cultivation of gardens at landscape scale as well as in more domestic contexts has bequeathed some areas of valued scenery to the present day. On the other hand, these protective attitudes were counterbalanced by environmental degradation in which little notice was taken of the deleterious effects of, for example, smoke pollution in cities (prevalent in London from at least the thirteenth century onwards) or of the ways in which lead mining in the uplands might kill not only the workers in the industry but the entire biota of runoff streams. Again, the slugs of lead fines in river alluvial deposits still kill off marginal plant life (including crops) at times of flood[1].

The coming of economies powered not only by the sun but also by fossil fuels (this meant predominantly coal in the case of the UK until the 1960s) induced many environmental changes, since the developing technologies meant, firstly, the concentration of many phenomena (as cities, as industrial complexes, as high-nitrogen chemicals for example) and, secondly, the emission of substances unknown in nature and hence having no evolved degradational pathways. Many organic compounds, metal alloys and pharmaceuticals are among such instances. All these were part of an economic anabasis of which a rapidly growing human population was an essential driving force. In response, a number of important legislative steps were taken in the period from ca 1850 to recent times. In some instances these lagged behind the tenor of the times and in a few they were pathbreaking; all were the result of the tensions between people, Parliament and, with the exception of the period 1945-1978, an economy in which state ownership of major industries was never dominant[2].

1 *I.G. Simmons*, An Environmental History of Great Britain from 10,000 Years Ago to the Present, Edinburgh 2001.

2 *J. Winter*, Secure from Rash Assault. Sustaining the Victorian Environment, Berkeley, Los Angeles and London 1999.

In the 19th century, however, the growing heavy industries of the UK were not accustomed to external control. Their emissions contaminated air, water and land alike, with the worst period of all coming towards the end of the 19th century. In 1881-5, central London in winter enjoyed only about one-sixth of the sunshine of southern small towns like Oxford and Cambridge. An industrial suburb of Sheffield in south Yorkshire at about the same time suffered the deposit of 55 tons per square mile (215 kg/ha) of solids, compared with the spa town of Malvern at 5 tons per square mile (19.6 kg/ha), much of which was earth and leaves and not the by-product of the burning of coal or other industrial particulates. There were attempts to control both smoke and chemical effluvia by means of Acts of Parliament and local authority bye-laws from 1821 onwards but most either failed to keep up with industrial processing techniques or were not enforced with any efficiency: a case in point is the Alkali Act of 1863 which was not fully effective until 1881 when the Alkali Works etc Regulation Act signalled by its "etc" that a broad range of industries were to be made to conform, albeit to standards that were not very high. Lower levels of smoke in cities from domestic fires had to wait until the aftermath of the Great London Smog of 1952 (see below) before proper regulation had much effect. In his account of the progress of the legislation in the later 19th century, B W Clapp emphasizes that the chief inspector (Dr R. Angus Smith, whose PhD of 1841 was obtained in Germany under the supervision of Justus von Liebig) appointed by the Board of Trade was a gentleman and thus able to talk persuasively to the industrialists: such was the role of the class structure in the UK at that time[3]. Neither he nor his successors could prevent the accumulation of contaminated land which even today, as a "brownfield" alternative for development (to spare the green fields of rural fringes of cities), presents problems of toxic load.

In its approach to a problem by fits and starts, the control of noxious substances had an analogy in the control of land use. The English in particular developed a tender attitude towards their countryside in the decades after the industrialization of cities and felt that most outward extensions of a city's suburbs contravened not only economics (for it was usually expensive for the local authority to supply services such as water and sewage) but a kind of moral law: ribbons of bungalows along the main roads might not exactly be Blake's "dark satanic mills", but they certainly took up pastures otherwise available to Holy Lambs. The phrase "urban sprawl" came to have very strong emotional overtones by the 1920s and so some local authorities attempted from 1909 onwards to enact bye-laws that corralled the developers[4]. However, most failed because there were no national scales of compensation that might be afforded to those who missed out on lucrative sales to the builders of houses, factories and new roads. So it was only in the wartime emergency atmosphere of 1943 that planning control for the whole country was emplaced, to be followed by the key

[3] *B.W. Clapp*, An Environmental History of Britain since the Industrial Revolution, Harlow 1994.
[4] *J. Sheail*, The Restriction of Ribbon Development Act: The Character and Perception of Land-Use Control in Inter-War Britain: Regional Studies 13, 1979, 501-512.

Town and Country Planning Act of 1947 (and its Scottish equivalent), one more feature of the tenure of Britain's most reform-minded government ever. Local authorities were given effective control of most developments, with the exceptions of farming and forestry, and were required to produce land-use plans, in effect what came to be known elsewhere as zoning. The Act hoped to find roles for both national and local government in an essentially rational and "scientific" approach to land use and it supported the ideas of public control of development rights and the containment of towns within "green belts". In many ways it was permissive and not all authorities have made use of all the powers all of the time: unwise development of housing and industry on flood-plains, for example, has been widespread in recent years[5]. Like all such legislation, local political influence and indeed outright central hostility have been common but they probably show how important and how good a waymarker it has been.

The public and academic attitudes to scenery and nature also shared in the inter-war concerns about development, with the thrust of public pressure directed towards a system of national parks and nature reserves. Despite the pressures from high-profile individuals who insisted that they were on the side of progress, modernity and not anti-urban, only in 1939 was a draft Bill prepared that addressed such questions of amenity and conservation. Unfortunately it lapsed in the face of war. However, once again, that conflict produced a national determination to make the best of the nation's resources and so in 1949 England and Wales (but not Scotland) were the subjects of the National Parks and Access to the Countryside Act. This allowed for the designation of ten areas (subsequently added to) of mostly upland terrain as National Parks, where planning control was stricter than elsewhere. Given that they were largely controlled by local authorities and that recreational access was subject to the prior demands of landowners, it was common to say that they were not national and certainly not parks. But the symbol was there: the jewels in the nation's scenic crown were not to be stolen and if possible polished a little. However, the greatest importance of the Act was arguably the setting up of the Nature Conservancy, with a remit to establish and manage National Nature Reserves and Sites of Special Scientific Interest and to conduct appropriate scientific research[6]. It also had powers of compulsory purchase which were scarcely ever used. The work of the Conservancy in protecting semi-natural habitats (though not always informed by accurate environmental histories) and forwarding knowledge of how to manage protected areas was world-class but it fell prey eventually to other landed interests and has been hacked about shamelessly by governments willing to appease the powerful[7]. In some fields, its obvious skills have yet to make their proper impact: the control of red deer num-

5 *J. Sheail*, An Environmental History of Twentieth-Century Britain, Basingstoke and New York 2001.

6 *G.E. Cherry*, Environmental Planning, vol II: National Parks and Recreation in the Countryside, London 1975. Scotland was included this time.

7 *J. Sheail*, Nature in Trust. A History of Nature Conservation in Britain, Glasgow 1976; idem., Nature Conservation in Britain: the Formative Years, London 1998.

bers in Scotland for example, is only just beginning to reflect the ecology of the Highlands as much as the iconography of Prince Albert surrounded by dead game. Equally, little progress has been made in peace-keeping between grouse moor managers and the predatory hen harrier[8].

Progress in improving air quality was slow once some impression had been made on factory chimneys and on chemical emissions into the atmosphere. Much of the problem was due to the British insistence on an open coal fire, which not only heated the air as well as the room but emitted smoke most of the time. In about 1751, the young Edward Gibbon was removed from Oxford to Lausanne where his small chamber, "*instead of a companionable fire must be warmed by the dull invisible heat of a stove*", an attitude prevalent in most of Britain until the late 1950s at least. In spite of attempts (once again) by some cities to impose local controls on the use of coal in domestic contexts, any diminutions in smoke during the period 1880-1920 were due to the lowering of urban population densities, some controls on industrial outputs, and slightly more efficient stoves and fireplaces. The turning point came in the capital. As with so many centralized states, if it does not happen in that city then it is scarcely worth noticing; however, London's smog of 1952 could not be ignored. Natural radiation-induced fog is a normal feature of the Thames Valley but the presence of a temperature inversion traps contaminants beneath it; the "London particulars" and "pea-soupers" beloved of Dickens and the makers of Sherlock Holmes movies were especially prevalent in the winters of the 19th century (the first recorded instance was by John Evelyn in 1671), followed by relatively infrequent occurrences during the often warmer winters of the 1900-1950 period. But from 5-9 December 1952, a cold fog occurred in which concentrations of 1.34 ppm of sulphur dioxide were recorded, about 10 times the normal. Visibility was rarely above 20m and often down to 10m (I remember it well since the train service to school was virtually non-existent for a few days) and on December 8, the audience in the Royal Festival Hall could not see the stage. However, the defining factor was the number of excess deaths, which was at least four thousand. The official report cut off the mortality totals after a few days and attributed them to "influenza", so that the true total is higher. Eventually, there ensued the Clean Air Act of 1956 which allowed local authorities to designate "smokeless zones" where the fuels and combustion devices in domestic premises (as well as industrial plants) could be regulated[9]. Not every town scrambled to take up the chance: some towns dominated by politicians in the thrall of coal-mining unions waited until the 1980s to introduce such zones. Of course, the Act was not comprehensive chemically either, since sulphur dioxide was not included and the later development of photochemical "smog" of the Los Angeles type is slow to attract any significant attempts at control, with the possible exception of the London

[8] *I.G. Simmons*, The Moorlands of England and Wales. An Environmental History 8000 BC to AD 2000, Edinburgh 2003, 244-247.

[9] *P. Brimblecombe*, The Big Smoke: A History of Air Pollution in London since Medieval Times, London and New York, 1987.

congestion charging zone[10]. There is, however, the paradoxical situation that air quality in London and some other big cities is now likely to be better in winter than in summer.

The story is told that in an early 20th century foggy period, a newspaper headline ran "*Fog in the Channel: Continent Isolated*". Hence, examples of Britain's engagement with a wider but post-imperial world are important signifiers of an acknowledgment of environmental interdependencies. Migratory waterfowl had for years engaged the attention of bird-watchers and sportsmen alike and the accession of the UK to the Ramsar Convention of 1971 on the protection of wetlands[11] is symbolic of a new era. That the main site of Ramsar status is The Wash, a mirror image of many of the landscape features of the coast of the Netherlands, Germany and Denmark, ties the UK into the wider environmental policies of the European Union generally[12]. Though the EU's major environmental influences are the Common Agricultural Policy and its Fisheries Policy (both for many years largely directed at production rather than protection), a number of directly environmental frameworks have been promulgated by the Commission. Examples include rare species and unique habitants, biodiversity generally, the supply and use of fresh water, and the quality of bathing beaches[13]. In a number of cases, the UK government has reneged on these agreements and has found itself in court for failing to keep them, perhaps indicating that the famed sense of isolation from "the rest of Europe" is still present. In an ironic sense, it is the marine environment which causes considerable problems: the fisheries policy may have forced lower catches from overfished areas like the North Sea but it has impacted differentially upon countries and communities in what seems an unfair way[14]. As might be expected, former common species like cod and herring are now becoming upmarket foods: cod in cafés and restaurants now tends to sport a "walnut and oatmeal crust with a timbale of wilted rocket" rather than batter; kippers feature on the expensive breakfast menu of first-class railway dining cars. Yet, the scientific knowledge of North Sea fish stocks and their dynamics is probably the best in the world[15]. This fragmentation between knowledge and political action is likely to become apparent when coastal management has to deal with higher sea-levels (and possible increases in storminess) in the near future. The responsibility for coastal geomorphology is divided amongst local authorities whose boundaries are not aligned to

10 This refers to the recent (2004) introduction of a £5 charge affecting most vehicular access to inner London.

11 In the words of the convention, "*wetlands are areas of marsh, fen, peatland or water, whether natural or artificial, permanent or temporary, with water that is static or flowing, fresh, brackish or salt, including areas of marine water the depth of which at low tide does not exceed six metres*".

12 *S. Frazier*, Ramsar Sites Overview. A Synopsis of the World's Wetlands of International Importance, Newbury 1999.

13 *D. Plut*, Environmental Challenges of Europe: Geojournal 52 (2000), 149-155.

14 J. *Beddington* and *G. Kirkwood* (eds.), Fisheries: a Future?: Philosophical Transactions of the Royal Society of London Biological Sciences 360 (2005), 3-218.

15 *J-P. Ducrotoy* and M. *Elliott*, Interaction Between Science And Policy-Making: the North Sea Example: Marine Pollution Bulletin 34 (1997), 686-701.

the dynamics of coastal cells of erosion and deposition which determine the likely fate of particular stretches, amended as they usually are by human activity[16]. A single reclaimed agricultural field in Lincolnshire whose sea-wall has been deliberately breached to "see what happens" is probably a portent no less vivid for being very near a prison.

These examples are selected as outstanding markers from the arena of official action. One interesting facet is the interaction between them and the evolution of the public consciousness of environment. Schools engage with environmental concerns at many levels until prevented from doing so by specialisation in the curricula. Nevertheless, school syllabuses in, for example, Geography, are sufficiently well tuned to environmental matters to incur attacks from neo-liberal media from time to time. Wildlife programmes and indeed environmental history series (some better than others) are a staple of the pre-2100 hrs slots on major TV channels. But to what extent the sight of a tiger on television enhances an individual's connections to the natural world better than pond-dipping is unknown. Delinquent youths are often sent on outdoor activities in mountain areas and doubtless they would vote to have such places concreted over. On the other hand, one major bird-protection organisation currently has more than one million members though many such groups are single-issue organisations and there is now some reluctance to get behind the more controversial groups like Greenpeace and Friends of the Earth. Equally, the environmental costs of food production and supply are concerns of a limited class of people.

Not long ago, it was common for environmental publications to imply that their era (say 1960-1975) was clearly the "turning point" in the development of a new consciousness towards the non-human world. Such claims are now less frequent, though arguments can be made for the success of neo-Malthusian attitudes towards population growth in the 1950s and 1960s. The questions of energy supply and the carbon economy's consequences now dominate the environmental agenda in most industrialized nations, leaving rather less room for nuanced environmental discussions, not least because the prediction of biotic response is so uncertain. Although in a much more demotic framework, the old attitudes of selective protection and "business as usual" outlined at the start of this essay are clearly still with us.

16 *S.J. Pettit*, The Statutory Approach to Coastal Defence in England and Wales: Marine Policy 23 (1999) 465-477.

Kai F. Hünemörder

Vom *„Pritzelkram"* zur *„europäischen Umweltpolitik"*: Schlüsselperioden des deutschen Naturschutzes vom Kaiserreich bis zur umweltpolitischen Wende am Beispiel Schleswig-Holsteins (1906-2004)

In den letzten Jahren hat die Aufmerksamkeit für die Geschichte des deutschen Naturschutzes deutlich zugenommen. Mittlerweile liegen sowohl für die Kaiserzeit als auch für die Epoche des Nationalsozialismus und die Nachkriegszeit historische Monographien und Sammelbände vor[1]. Da sich die Folgen der Wirtschaftstätigkeit und die Wirkungen der Naturschutzbewegung allerdings in der Kulturlandschaft vor Ort zeigen, gewinnt die Geschichte des *„Kampfes um die Natur"* nur durch regionale Tiefenschärfe weiter an Konturen. Gefragt sind regionale Fallbeispiele, an denen Einflüsse und Rückwirkungen nationaler und internationaler Prozesse aufgezeigt werden können. Besonders wenig erforscht ist in diesem Zusammenhang die Zeit des Nationalsozialismus. Frank Uekötter hat unlängst auf einen „eklatanten Mangel an Lokalstudien" hingewiesen, *„die dokumentieren, was die NS-Zeit ganz konkret* [...] *für eine Region bedeutete"*[2].

Da gerade der Norden Deutschlands in der bisherigen Forschung stiefmütterlich behandelt worden ist, werden im Folgenden drei Perioden des beschleunigten Wandels in Schleswig-Holstein skizziert[3]. Nach der Institutionalisierung des Naturschutzes vor dem ersten Weltkrieg werden insbesondere die Wandlungen des Agierens der

[1] *Friedemann Schmoll*, Erinnerung an die Natur. Die Geschichte des Naturschutzes im deutschen Kaiserreich, Frankfurt/Main u.a. 2004; *Andreas Knaut*, Zurück zur Natur! Die Wurzeln der Ökologiebewegung, Supplement 1 (1993) zum Jahrbuch für Naturschutz und Landschaftspflege, Bonn 1993; Naturschutz und Nationalsozialismus, hg. v. Joachim Radkau und Frank Uekötter, Frankfurt/Main u.a. 2003; *Frank Uekötter*, The Green and the Brown. A History of Conservation in Nazi Germany, Cambridge 2006; *Jens Ivo Engels*, Naturpolitik in der Bundesrepublik. Ideenwelt und politische Verhaltensstile von Naturschutz und Umweltbewegung 1950-1980, Paderborn 2006; *Kai F. Hünemörder*, Die Frühgeschichte der globalen Umweltkrise und die Formierung der deutschen Umweltpolitik, Stuttgart 2004.

[2] *Frank Uekötter*, Einleitung, in: Naturschutz und Nationalsozialismus, hg. v. Joachim Radkau und Frank Uekötter, Frankfurt/Main u.a. 2003, 26.

[3] Neben dem Bestand der Kieler Universitätsbibliothek zum Themengebiet „Naturschutz in Schleswig-Holstein" wurden einige Bestände in den Bundesarchiven in Koblenz und Berlin sowie im Landesarchiv in Schleswig ausgewertet.

Naturschutzbewegung im „Dritten Reich" und in den 1960er und 1970er Jahren bezogen auf den Inselnaturschutz in den Blick genommen. Zum Ausklang wird die Wirkung der Europäischen Union als Akteur im Naturschutz diskutiert.

Noch immer wird der frühe Naturschutz von vielen einseitig mit rückwärtsgewandter Modernisierungskritik assoziiert. Als Beleg wird dabei häufig an den Ausspruch des bekannten Heidedichters und zivilisationskritischen Naturschützers Hermann Löns von 1911 vor dem Bremer Lehrerverein erinnert: *„Pritzelkram ist der Naturschutz, so wie wir ihn haben. Der Naturverhunzung dagegen kann man eine geniale Großzügigkeit nicht absprechen. Sie fährt Auto im 80 Kilometertempo; der Naturschutz kraucht knickebeinig hinterdrein. Die Naturverhunzung arbeitet ‚en gros'; der Naturschutz ‚en detail'. Die Naturverhunzung herrscht, der Naturschutz steht zu ihren Diensten"*[4].

Der einseitige Fokus auf die radikalen Kritiker an der Naturzerstörung seit der Hochindustrialisierung verstellt den Blick auf die generelle Nähe eines großen Teils der frühen Naturschutzbewegung zum wilhelminischen Staat. Joachim Radkau hat zu Recht darauf hingewiesen, dass der Naturschutz *„zu Wissenschaft, Wirtschaft und Technik sogar bemerkenswert gute Beziehungen"* unterhielt. *„Und er besaß eine Affinität zu vielen Reformbewegungen jener Zeit: dem Wandervogel, der Reformpädagogik und dem breiten Spektrum der Hygiene-, Gesundheits- und Lebensreformer, die untereinander in einem dynamischen Zusammenhang standen"*[5].

Das dominante Bewahrkonzept des frühen Naturschutzes konnte sich bald auf einen weitgehenden gesellschaftlichen Konsens stützen. Es bestand darin, herausragende Relikte aus früheren Zeiten zu schützen. Wie Friedemann Schmoll gezeigt hat, ging es dabei in einer Zeit beschleunigten Wandels mental *„um die Erinnerung an eine durch den historischen Wandel gefährdete natürliche Umwelt"*. Eine *„denkmalhaft zu rettenden Natur"* suggerierte in der hektischen Zeit der Hochindustrialisierung Kontinuität. Wie stark sich der frühe Naturschutz nicht nur auf der Ebene der 1906 gegründeten Staatlichen Stelle für Naturdenkmalpflege in die Politik des Kaiserreichs einfügte, soll kurz am Beispiel der Gründung des Schleswig-Holsteinischen Provinzialausschusses für Naturdenkmalpflege und des Schleswig-Holsteinischen Landesvereins für Heimatschutz erläutert werden. Die Zeit unmittelbar vor dem ersten Weltkrieg lässt sich als erste Schlüsselperiode für die Geschichte des deutschen Naturschutzes begreifen.

[4] *Herrmann Löns*, Der Naturschutz und die Naturschutzphrase. Ein noch unbekannter Kampfruf von Herrmann Löns, in: Der Waldfreund 5 (1929), 4 f., zitiert nach *Schmoll* (wie Anm. 1), 97.

[5] *Joachim Radkau*, Natur und Macht. Eine Weltgeschichte der Umwelt, München 2000, 271. – Zu den einzelnen Reformbewegungen siehe Handbuch der deutschen Reformbewegungen, hg. v. *Diethart Kerbs* und *Jürgen Reulecke*, Wuppertal 1998.

I. Naturschutz vor dem 1. Weltkrieg und in der Weimarer Republik: Zur Formierungsphase der Schleswig-Holsteinischen Naturschutzbewegung (1906-1933)

Der Naturschutz als gesellschaftliche Aufgabe entstand auch in Schleswig-Holstein zu Beginn des 20. Jahrhunderts auf der Grundlage einer älteren Tradition. So gab es bereits im 18. Jahrhundert Bemühungen, die Dünen im Norden der Insel Sylt zu schützen. Die Verordnung aus dem Jahre 1739 hatte freilich nicht den Schutz der Natur um ihrer selbst willen zum Gegenstand, sondern zielte auf den Schutz der Insel gegenüber Sturmfluten. Mit Blick auf die Kosten künstlicher Deiche schätzten schon die Zeitgenossen die Funktion natürlicher Dünen[6]. Auch sonst blieben Initiativen, die nicht mit direkten Nutzungs- oder naturwissenschaftlichen Erkenntnisinteressen zusammenhingen, vor der Entstehung der Naturschutzprogrammatik vereinzelt. Zu nennen ist etwa die Gründung des Naturwissenschaftlichen Vereins für Schleswig-Holstein, in dessen Reihen frühzeitig Schutzüberlegungen angestellt wurden. Wie in anderen Regionen erwiesen sich die Ornithologen, die 1876 unter dem Vorsitz des Husumer Gymnasiallehrers Joachim Rohwedder eine eigene Sektion bildeten, als besonders sensibel gegenüber den Veränderungen im Artenbestand[7].

Bald nach Gründung der erwähnten Staatlichen Stelle für Naturdenkmalpflege initiierte ihr Leiter, der preußische Geheimrat Hugo Conwentz, die Bildung regionaler Untergliederungen[8]. Mit Hilfe von Angehörigen der staatlichen Elite in Kiel und Schleswig wurde im März 1909 das Schleswig-Holsteinische Provinzialkomitee für Naturdenkmalpflege geschaffen[9]. Vorbild für die Gründung korrespondierender Komitees in den einzelnen Provinzen war die Gliederung des preußischen Denkmalschutzes. Zu Beginn hatte das Naturschutzkomitee etwa fünfzig Mitglieder, vor allem höhere Regierungs- und Kommunalbeamte, Professoren und naturkundlich interessierte Lehrer. Zugleich wurde ein siebenköpfiger Arbeitsausschuss eingesetzt, den die Kieler Professoren dominierten[10]. Mit der Geschäftsführung wurde der Alto-

6 Vgl. *Johann Georg Kohl*, Die Marschen und Inseln der Herzogtümer Schleswig und Holstein, 3 Bde., Dresden und Leipzig 1846, passim.

7 Die Gründung des Naturwissenschaftlichen Vereins im Jahr 1855 – zunächst unter dem Namen *Verein nördlich der Elbe zur Verbreitung Naturwissenschaftlicher Kenntnisse* – geht vor allem auf das Abgrenzungsbedürfnis der Deutschen gegenüber Dänemark in der nationalpolitischen Auseinandersetzung zurück. Vgl. etwa *Ekkehard Schmidt*, 30 Jahre LNV – 150 Jahre Naturschutz in Schleswig-Holstein, hg. v. Landesnaturschutzverband Schleswig-Holstein e.V. (LNV), Kiel 2005, 13 ff.

8 Zur Staatlichen Stelle siehe *Michael Wettengel*, Staat und Naturschutz 1906-1945. Zur Geschichte der Staatlichen Stelle für Naturdenkmalpflege in Preußen und der Reichsstelle für Naturschutz, in: *HZ* 257 (1993), 355-399.

9 Wie in den meisten anderen Provinzen übernahmen Honoratioren wie der Vorsitzende des Schleswig-Holsteinischen Provinziallandtages, Graf von Reventlou-Damp, und Oberpräsident von Bülow den Vorsitz. Bei der Eröffnungssitzung im Auditorium maximum der Kieler Universität hielt Conwentz vor etwa 130 Personen einen Lichtbildvortrag über die Erhaltung der Naturdenkmäler, in dem er die Verhältnisse in der Provinz besonders beleuchtete. Vgl. Landesarchiv Schleswig-Holstein (*LA Schleswig*) Abt. 301 Nr. 1192.

10 Eine starke Position nahm dabei anfangs der Zoologe Brandt ein. *LA Schleswig* Abt. 301 Nr. 1192, Schreiben von Conwentz an den Oberpräsidenten Herrn v. Bülow v. 30.1.1909 und

naer Oberlehrer Wilhelm Heering betraut, der sich zuvor durch die Veröffentlichung eines Forstbotanischen Merkbuches für die Provinz Schleswig-Holstein für diesen ehrenamtlichen Posten empfohlen hatte.[11]

Bereits im Vorjahr gelang es auch dem Bund Heimatschutz in Schleswig-Holstein Fuß zu fassen. Obwohl der Initiator und erste Vorsitzende des Landesvereins für Heimatschutz, der Gutsbesitzer Paul von Hedemann-Heespen, *„Landschaft und* [...] *die Ortschaften in Schleswig-Holstein bisher* [...] *weniger durch unzeitliche Anlagen verunstaltet* [*sah*] *als im deutschen Reiche"*[12], formulierte er insbesondere im baulichen Bereich Handlungsbedarf. Der Landesverein, der sich wie das Provinzialkomitee trotz seiner Orientierung auf breitere Bevölkerungsschichten zu Beginn stark auf Honoratioren, besonders Beamte und Architekten stützte, setzte sich das Ziel, *„die Eigenart und die Denkmäler der Heimat zu schützen, zu pflegen und fortzubilden, besonders in Landschaft und Bauweise, Kunstgewerbe, Gebräuchen und Trachten, Sprache und Namen, Tier- und Pflanzenwelt und in den Denkmälern aus der Vorzeit der Natur und der Kultur"*[13].

Für beide neuen Organisationen nahm der lokale und regionale Bezug eine bedeutende Funktion ein. In anschaulichen Lichtbildervorträgen über die Heimatnatur und die Heimatarchitektur warb man im ganzen Land um Mitglieder und Spenden. Auch aufgrund eines relativ niedrigen Mindestbeitrags von drei Mark wuchs die Zahl der Mitglieder des Heimatschutzvereins bis 1914 auf etwa 270 an. Daneben traten mehrere Vereine und Behörden bei. Bald wurden im Einklang mit den Vorstellungen im Oberpräsidium mehrere Ortsgruppen ins Leben gerufen. In ihnen waren bis 1914 über 500 Schleswig-Holsteiner organisiert[14].

Abschrift des Protokolls v. 22.3.1909 samt Anlagen mit einer Mitgliederliste. Rechtliche Grundlage für die Gründung war der Erlass des Kultusministers v. 30.5.1907 betr. Grundsätze für die Förderung der Naturdenkmalpflege in den Provinzen. *LA Schleswig* Abt. 301 Nr. 4066, Schreiben von Emeis an den Oberpräsidenten v. 7.6.1928. – Zur Übersicht über die Anfänge in den anderen preußischen Provinzen siehe *LA Schleswig* Abt. 301 Nr. 1192, Bericht über die Staatliche Stelle für Naturdenkmalpflege in Preußen im Jahre 1906, Beiträge zur Naturdenkmalpflege, hg. v. *H. Conwentz*, Berlin 1907.

[11] Forstbotanische Merkbücher sollten die *„beachtenswerten und zu schützenden urwüchsigen Sträucher, Bäume und Bestände"* nachweisen. *Wilhelm Heering*, Bäume und Wälder Schleswig-Holsteins. Ein Beitrag zur Natur- und Kulturgeschichte der Provinz. Im Auftrage des Naturwissenschaftlichen Vereins für Schleswig-Holstein bearbeitet, Kiel 1906, 4. Siehe auch *ders.*, Forstbotanisches Merkbuch. Nachweis der beachtenswerten und zu schützenden urwüchsigen Sträucher, Bäume und Bestände im Königreich Preußen. IV. Provinz Schleswig-Holstein, Berlin 1906.

[12] Die Gründung des Landesvereins erfolgte nach intensiven Vorbereitungen des Oberpräsidiums am 15. Oktober 1908 im Haus der Landwirte in Kiel. Hedemann-Heespen trat auch dem Provinzialkomitee bei. Zur Vorgeschichte der Gründung und den Anregungen des Bundes Heimatschutz an den Oberpräsidenten v. Bülow siehe die die diversen Briefe in *LA Schleswig* Abt. 301 Nr. 1211 und Kieler Zeitung v. 16.10.1908 („Landesverein des Bundes Heimatschutz").

[13] In: Satzungen des Schleswig-Holsteinischen Landesvereins für Heimatschutz. Eingetragener Verein. Angenommen in der Gründungsversammlung am 15. Oktober 1908. Vgl. auch *LA Schleswig* Abt. 301 Nr. 1211, Schreiben von Hedemann-Heespen an den Kreisausschuss des Kreises Eiderstedt v. 22.12.1908.

[14] Die ersten 4 Ortsgruppen bildeten sich in Lauenburg, Neumünster, den Elbgemeinden und in Husum. Vgl. *Sechs Jahre Arbeit des Schleswig-Holsteinischen Landesvereins für Heimatschutz 1908-1914*, im Auftrage des Vorstandes von Paul v. Hedemann-Heespen, 3 und *LA Schleswig*

Wie in anderen Landesvereinen wurde die umfangreiche Programmatik des geistigen Vaters und Mitbegründers des Bundes Heimatschutz, Ernst Rudorff, zur umfassenden kulturellen Erneuerung in Schleswig-Holstein nur reduziert aufgegriffen. Es kam in den Folgejahren nicht zu einer für den Schutz von Natur, Landschaft und Kultur gleichermaßen kompetenten Heimatschutzbewegung[15]. So wurde die Arbeit dominiert vom Versuch der Rückkehr zu landestypischen Baustilen[16].

Ein Mittel zur Verbreitung einer Vorstellung vom *„Bild der überlieferten Heimat in Natur und Kultur"* war die Sammlung und Vorführung von Diapositiven. 1909 wurde die Lichtbildersammelstelle, die vorher im Kieler Thaulowmuseum untergebracht war, dem Landesverein angegliedert. Direkteren Einfluss versuchte man auf die Baubehörden und die Baupolizei zu gewinnen. Die gesetzlichen Grundlagen beförderten zunächst jedoch nicht eine wirksame Verbreitung heimatlicher Baustile[17].

Als elementar für das schleswig-holsteinische *„Landschaftsgefühl"* erachtete man auch die *„Hünengräber der germanischen Vergangenheit"* und die *„Waldemarsmauer des Danewerks"*[18]. Trotz des architektonischen Schwerpunkts befassten sich einzelne Ortsgruppen wie die Husumer Heimatschützer von Beginn an besonders mit Fragen des Landschaftsschutzes[19]. In der Tradition Rudorffs kritisierte der Heimatschutzbund etwas später auch im Norden die Auswirkungen der Meliorations-

Abt. 301 Nr. 1211, Dritter Jahresbericht der Ortsgruppe Heimatschutz für den Kreis Herzogtum Lauenburg und Umgebung, Ratzeburg 1914.

[15] Vgl. *Wolfgang Erz*, Geschichte des Naturschutzes. Rückblicke und Einblicke in die Naturschutz-Geschichte, in: Natur und Landschaft: Zeitschrift für Naturschutz und Landschaftspflege 65 (1990) 3, 103-106, hier 104.

[16] Von Beginn an erachtete der Vorstand die Gestaltung der Dächer als entscheidend für die Rettung des Bildes der Ortschaften. Gegenüber den neueren Zementdächern und Flachdächern aus Pappe und Wellblech versuchte man den Rückzug der Reetdächer zu verhindern. Der konzertierte Einsatz gegen Blechbedachungen stieß bald auf scharfe Gegenwehr des Vereins deutscher Eisenhüttenleute. Mit einer Denkschrift wandte er sich 1915 an höchster Stelle gegen die *„Auswüchse der Heimatschutzbewegung"* und ihren *„verhängnisvollen Einfluß auf die Behörden"*. Nach Ansicht von Hedemann-Heespen konnte man sich angesichts des überstürzten Städtebaus und *„einer retdachfeindlichen Versicherungspolitik"* nur noch in zurückgebliebenen Winkeln des Landes *„ein Bild von den Zeiten anständigen Geschmacks"* machen. Gemeinsam mit dem Regierungsbaumeister vom Kaiserlichen Kanalamt, Dr. Charton, und dem Kieler Baurat Carl Meyer versuchte Hedemann-Heespen einer historisierenden Bauweise zum Durchbruch zu verhelfen, indem man eine Bauberatungsstelle schuf und einen *„Baukatechismus"* mit Mustervorlagen verbreitete. Zitate aus: *LA Schleswig* Abt. 301 Nr. 1211, u.a. Denkschrift betr. Auswüchse der Heimatschutzbewegung: Verfolgung der Blechbedachung und Anschreiben an den Minister des Innern v. 30.6.1915, *Paul von Hedemann-Heespen*, Die Herzogtümer Schleswig und Holstein in der Neuzeit, o.O. 1926, 904. Für eine frühere Flugschrift mit Regeln für den Hausbau siehe auch die Broschüre des Heimatschutzvereins *„An die Bauherren und Baugewerksmeister der Landorte"*. Zu Baumanpflanzungen heißt es: *„In die Umgebung pflanze man schöne Bäume, Linden, Ahorn, Eichen, Ulmen, hochragende Pappeln. Einen prächtigen Schmuck für ein Landhaus bietet eine Reihe streng gezogener und beschnittener Linden vor der Front."* Vgl. auch *Sechs Jahre Arbeit* (wie Anm. 14), 10 f.

[17] Der frühe Heimatschutz orientierte sich an dem so genannten *„Zopfstil"*, wie er in einigen Städten und Gütern noch zum Ausdruck kam. Vgl. ebd., 8.

[18] *Hedemann-Heespen* (wie Anm. 16), 912 f., *LA Schleswig* Abt. 301 Nr. 1211, Dr. Lehmann, Das Danewerk (Vortrag mit Lichtbildern), um 1907.

[19] *Sechs Jahre Arbeit* (wie Anm. 14), 5.

maßnamen des 19. und frühen 20. Jahrhunderts. Durch Dränage und Entwässerung seien mit dem *„schwarzen Storch"* der *„König der Wälder"* und mit dem Kranich *„der König der Retfelder"* verschwunden. Anfang der 1930er sah man auch den *„weißen Storch"* als *„König der Weiden und Wiesen"*[20] als gefährdet an.

Wie aber sah es im Kaiserreich mit den rechtlichen Möglichkeiten für den Artenschutz aus? Über das Reichsvogelschutzgesetz von 1888 hinaus konnte sich die frühe Naturschutzbewegung nicht auf scharfe legislative Waffen stützen. So blieb im Wesentlichen nur die Möglichkeit, über den Ankauf und die Betreuung relevanter Flächen etwa den Schutz der Vogelwelt an der deutschen Nordseeküste voranzubringen. In diesem Sinne konzentrierte sich der *Verein Jordsand* zu Beginn seiner Tätigkeit auf die Einrichtung von Seevogelfreistätten und den Brutvogelschutz gegenüber *„Eierräubern"*. Nach dem Ankauf des mittlerweile weggespülten Jordsandes und der *„Insel"* Norderoog, den neben den Vereinsmitgliedern die Begründerin des Bundes für Vogelschutz, Lina Hähnle, mit einem zinsfreien Darlehen unterstützte, pachtete der Verein auch den vogelreichen Ellenbogen auf Sylt[21]. Anhand dieses Gebietes lässt sich das Ende der ersten Schlüsselperiode des deutschen Naturschutzes aufzeigen. Mit Beginn des ersten Weltkrieges verloren die Naturschutzvereine nicht nur eine große Zahl ihrer Mitglieder[22]. Zugleich wurde ihr Handlungsrahmen durch die Militärbehörden deutlich eingeschränkt. So war der Ellenbogen für mehrere Jahre *„militärisch besetzt"*[23], so dass die Entsendung eines Wärters unterbleiben musste. Im Jahre 1919 konstatierte der Vereinsvorsitzende Franz Dietrich: *„Auf dem Ellenbogen ist nach den von mir eingezogenen Erkundigungen von dem früheren, im letzten Friedensjahre so aussichtsreich entwickelten Vogelleben nichts mehr übrig geblieben"*[24].

Im Inland litten besonders die Moore an der Vermehrung der Anbauflächen zur Erhöhung der Erntemengen im 1. Weltkrieg. Bereits Ende 1915 soll das Gesamtareal der Meliorierungsgebiete in der Provinz Schleswig-Holstein etwa 24.000 Hektar betragen haben. Funktionäre der Naturschutzbewegung gaben nicht der deutschen Seite, sondern der *„englischen Aushungerungspolitik"* öffentlich die Schuld für den Umschwung, der das Deutsche Reich gezwungen habe, *„sich mit seiner Volksernäh-*

[20] *Paul v. Heedemann-Heespen*, Wasser. Vortrag, gehalten auf der 2. Arbeitstagung Dithmarscher Heimatforscher, Sonderdruck aus der Zeitschrift Dithmarschen. Blätter für Heimatpflege und Heimatkultur, März-April-Heft (1932), 34-40, hier 40. – Zur Verbreitung des Weißstorchs im nördlichen Landesteil in den 1920er und 30er Jahren siehe *LA Schleswig* Abt. 301 Nr. 4066, *Walther Emeis*, Die Verbreitung des Weißen Storches als Brutvogel im heutigen Schleswig, Sonderdruck aus Schriften des Naturwissenschaftlichen Vereins für Schleswig-Holstein 17 (1926) 2, 256-262, und *LA Schleswig* Abt. 301 Nr. 4065, *Walther Emeis*, Die gegenwärtige Verbreitung des Weißen Storches in Schleswig-Holstein, Sonderabdruck aus Die Heimat 5 (1931).

[21] Vgl. Sonder-Abdruck aus der Ornithologischen Monatsschrift XXXV, Nr. 1, 42-56, hier 43.

[22] Der Verein Jordsand verlor von 1914 bis 1918 etwa 40% seiner Mitglieder. Schreiben vom Mai 1918, in: *LA Schleswig* Abt. 320 Eiderstedt Nr. 1806, Akten betreffend den Verein Jordsand zur Begründung von Vogelfreistätten an den deutschen Küsten.

[23] Vorläufiger Bericht über die Vereinstätigkeit im Jahre 1915, in: Ornithologische Monatsschrift XLI, Nr. 5, 161-163, hier 163.

[24] *Franz Dietrich*, Bericht über die Brutergebnisse des Jahres 1919, in: Ornithologische Monatsschrift XLV, Nr. 1, 6-24, hier 11.

rung auf eigene Füße zu stellen"[25]. Für die schweren Arbeiten wurden zunächst zahlreiche russische Kriegsgefangene eingesetzt, die nach dem Krieg durch schleswig-holsteinische Strafgefangene ersetzt wurden[26]. Noch während des Krieges setzten sich die Naturschützer in Berlin und Holstein dafür ein, Moorschutzflächen innerhalb der Kultivierungsgebiete zu schaffen[27].

Trotz Aufnahme des Naturschutzes in die Reichsverfassung änderte sich in der Weimarer Republik an den rechtlichen Möglichkeiten des norddeutschen Naturschutzes zunächst wenig. Einige Gebiete wurden zwar durch Polizeiverordnungen nach § 34 des Feld- und Forstpolizeigesetzes unter Naturschutz gestellt. Ministerielle Entwürfe für ein preußisches Naturschutzgesetz konnten sich allerdings nicht durchsetzen[28]. Im Inselnaturschutz wirkten sich vor allem die ökonomischen Krisenzeiten negativ aus. So wurden die Vogelschutzbemühungen auf der Nordseeinsel Trischen durch den *„Raub"* von Seevögeleiern Ende der 1920er Jahre deutlich erschwert. Denn zwischen Silbermöwen- und Seeschwalbennestern unterschieden die Menschen nur, wenn sie an Naturschutz Interesse zeigten[29].

Ein gemeinsames Problem der Nordseeanrainer war das Massensterben von Seevögeln. In den frühen 1920er Jahren breitete sich die Ölverschmutzung schnell aus, da die Handels- und Kriegsflotten verstärkt von Kohle auf Ölfeuerung umgestellt wurden[30]. Der Protest regte sich nicht nur bei wirtschaftlich interessierten Gruppen

[25] Mitteilungen des Schleswig-Holsteinischen Provinzialkomitees für Naturdenkmalpflege, Nr. 6 (1918), 14 f., in: *LA Schleswig* Abt. 301 Nr. 1193.

[26] Siehe *LA Schleswig* Abt. 309 Nr. 33780.

[27] Vgl. *LA Schleswig* Abt. 301 Nr. 1193 und Sicherung von Naturdenkmälern bei der bevorstehenden Kultivierung der Ödländereien, Denkschrift der Staatlichen Stelle für Naturdenkmalpflege in Preußen, Berlin 1919.

[28] Vgl. für die Initiativen *LA Schleswig* Abt. 301 Nr. 4066. Siehe auch Anweisung zur Ausführung des Gesetzes vom 8. Juli 1920 betr. die Abänderung des § 34 des Feld- und Forstpolizeigesetzes vom 1. April 1880, Sonderdruck aus dem Deutschen Reichsanzeiger und Preußischen Staatsanzeiger Nr. 1 v. 3.1.1921, in: *LA Schleswig* Abt. 301 Nr. 1193.

[29] Das in vielen Gebieten übliche Sammeln von Eiern der Möwen und des Kiebitz war aufgrund des § 42 der Jagdordnung in den 1910er und 20er Jahren für Jagdberechtigte bis zum 30. April erlaubt. Seit den Kriegsjahren 1915/16 wurde die Frist *„aus Gründen der hohe[n] Bedeutung der Möveneier für die Volksernährung"* regelmäßig bis zum 20. Mai verlängert. In einigen Gebieten wie auf der Insel Trischen war das Recht des Eiersammelns verpachtet. Die Naturschützer tolerierten dies, da ihrer Ansicht nach besonders das wilde Sammeln von Badegästen die Brutkolonien anderer Vogelarten wie der Seeschwalben störte. Aus wirtschaftlichen Interessen achteten die Pächter besser auf die Kolonien, als es staatlicher Schutz ohne polizeiliche Bewehrung vermocht hätte; denn nach 1914 wurden im Sommer keine Vogelwärter mehr auf der Insel zu Kontrollzwecken stationiert. Erst ab 1932 wurde wieder ein Wärter des Bundes für Vogelschutz zur Bewachung von Trischen eingesetzt. Im Vorjahr waren wiederholt Versuche von Büsumer Fischern und Badegästen beobachtet worden, Vogeleier zu sammeln. Der Wärter führte zugleich naturschutzfachliche Beobachtungen durch. *LA Schleswig* Abt. 309 Nr. 12287, Schreiben v. Stellvertretenden Generalkommando an den Regierungspräsidenten v. 20.4.1917, Schreiben von Richelsen an Hansen v. 21.4.1917 u. Schreiben v. Vorsitzenden des Kreisausschusses an den Regierungspräsidenten v. 26.4.1917; Bundesarchiv Koblenz (*BA Koblenz*) 245-64, Schreiben von R. Wendehorst an Ministerialdirektor Arnoldi v. 1.1.1933; *BA Koblenz* B 245-66, fol. 102 und *BA Koblenz* B 245-95, fol. 94.

[30] Laut Anna-Katharina Wöbse stieg der Anteil der ölbefeuerten Schiffe von 2,6% im Jahr 1914 bis 1922 bereits auf 22,3% an. Vgl. *Anna-Katharina Wöbse*, Der Schutz der Natur im Völkerbund – Anfänge einer Weltumweltpolitik, in: AFS 43 (2003), 177-190, hier 185.

wie Fischern, Hafenbesitzern, Küstengemeinden und Hotelbesitzern[31]. Zu den *„Entdeckern"* der Ursachen des Massensterbens gehörten auch Naturschützer wie der Leiter der Biologischen Station Helgolands, Hugo Weigold. Seit dem Spätsommer 1919 fand er an den Küsten Helgolands eine Reihe siecher, verendender oder bereits toter Vögel und schlug im Jahrbuch des Bundes für Vogelschutz Alarm: *„Wir erleben in der ganzen Nordsee ein Massensterben [von Seevögeln], das ganz neuartig ist. Selbst die lebhafteste Phantasie wird nicht darauf kommen, was diesmal eine so ernste Gefahr für den Helgoländer [Vogelbestand] ausmacht* [...] *Alle diese Vögel haben ihr Gefieder mit einem teerartigen Stoff verschmiert.* [...] *Was es ist, habe ich noch nicht ermitteln können"*[32].

Um die Ursache zu finden, schrieb Weigold an mehrere europäische Ornithologenkollegen, darunter auch einige Briten. In England nannte man das Phänomen bereits *„the Oil-menace"* oder *„oil-pollution"*. Die Folgen für die Avifauna wurden frühzeitig auch auf der benachbarten staatlichen Vogelwarte Helgoland registriert. Ihr zweiter Leiter Rudolf Drost soll – angesichts des Ausmaßes der Ölverschmutzungen vor der einzigen deutschen Hochseeinsel – zusammen mit Weigold 1928 den Begriff *„Ölpest"* geprägt haben. Er sah in der Häufung verölter Vögel eine Folge der vielen im 1. Weltkrieg versenkten Dampfer[33]. Dazu kam die gängige Praxis, Rest- und Abfallöl in Küstennähe und auf See gezielt abzulassen. Die internationalen Ansätze zur Eindämmung dieses grenzüberschreitenden Problems durch den Völkerbund scheiterten mit Beginn des Zweiten Weltkrieges[34]. Der Einbau von Ölabscheidern wurde nicht zur Pflicht erklärt. Trotz steigenden Problemdrucks konnte sich selbst ein internationales Einleitungsverbot innerhalb einer 50 Seemeilen-Zone zum Schutz der Küsten nicht durchsetzen[35].

II. Zum Naturschutz im „Dritten Reich" und im 2. Weltkrieg: Zwischen neuen gesetzlichen Grundlagen und den zerstörerischen Wirkungen der NS-Kriegspolitik in Schleswig-Holstein (1933-1945)

Mit der Machtübergabe an die Nationalsozialisten schienen sich für viele Beobachter die Handlungsspielräume der Naturschützer grundlegend zu ändern. Blickt man auf die gesteigerten Aktivitäten der Naturschützer und die beschleunigte Ausweisung

[31] Vgl. *Dies.*, Grenzfälle – Internationaler Naturschutz im 20. Jahrhundert (1. Teil), in: Studienarchiv Umweltgeschichte, hg. v. Institut für Umweltgeschichte und Regionalentwicklung e.V. und dem Förderverein der Stiftung Archiv, Forum und Museum zur Geschichte des Naturschutzes in Deutschland 8 (2003), 4-14, hier 12.

[32] *Hugo Weigold*, Das große Lummensterben, in: Jahrbuch des Bundes für Vogelschutz 1920, 149-152, zitiert nach *Wöbse* (wie Anm. 30), 184.

[33] Vgl. *Ekkehard Schmidt*, Handbuch für Naturschutz in Schleswig-Holstein, o.O. o.J., 106. – Siehe auch *Rudolf Drost*, Die Oelpest und ihre Wirkung auf die Lebewelt des Meeres, in: Der Naturforscher 4 (1927/28) 11, 540-544.

[34] Vgl. *Wöbse* (wie Anm. 31), 13.

[35] Vgl. *Dies.* (wie Anm. 30), 186-189.

von Naturschutzgebieten, kann die NS-Zeit für Schleswig-Holstein als zweite Schlüsselperiode angesehen werden.

Der Direktor der Staatlichen Stelle für Naturdenkmalpflege, Walther Schoenichen, bemühte sich seit 1933 explizit *„das in den Vordergrund zu stellen, was Nationalsozialismus und Naturschutz miteinander in engere Beziehung setzt"* und verknüpfte den Naturschutzgedanken programmatisch mit dem der *„Volksgemeinschaft"*. Naturverbundenheit sollte zudem *„Merkmal unserer Rasse"* bleiben: *„So ist es für uns eine unabweisbare Pflicht, die deutschen Gaue den kommenden Geschlechtern so zu überliefern, daß auch sie darin sich heimisch fühlen und daß auch sie im Umgang mit dieser Landschaft die deutschgemäße Rasseprägung ihres Wesens immer aufs neue zu vertiefen und zu verinnigen vermögen"*[36].

Während die Gleichschaltung aller Naturschutzvereine im *„Reichsbund Volkstum und Heimat"* scheiterte, wurden sozialdemokratisch orientierte Vereine wie der Touristenverein *„Die Naturfreunde"* zerschlagen[37]. Die massiven Ideologisierungs- und Zentralisierungstendenzen schlugen sich auch auf die Naturschutzprioritäten im waldarmen Schleswig-Holstein nieder. Der Schutz des *„deutschen Waldes"* stand nun im Vordergrund, hatte doch Wilhelm Heinrich Riehl bereits Mitte des 19. Jahrhunderts die Bedeutung des Waldes als *„Kraftquelle"* hervorgehoben und gefordert: *„Wir müssen den Wald erhalten, nicht bloß damit uns der Ofen im Winter nicht kalt werde, sondern auch damit die Pulse des Volkslebens warm und fröhlich weiter schlagen, damit Deutschland deutsch bleibe"*[38]. Bereits im Januar 1934 wurde ein Gesetz zum Schutz des Waldes gegen *„Waldverwüstungen"* erlassen, wonach Kahlschläge und Rodungen in Privatwäldern verboten wurden. 1936 wurden im Einklang mit der besonderen Wertschätzung der Wallhecken mehrere Verordnungen zur Erhaltung der Knicks etwa im Lübecker Raum erlassen[39]. Um die Vielzahl der folgenden Gesetze bekannt zu machen, gab man 1937 eine Sammlung der *„forst- und holzwirtschaftlichen Gesetzgebung des Dritten Reiches"* heraus[40].

Belege dafür, dass die norddeutschen Naturschutzfunktionäre bereits 1933 den Nationalsozialismus euphorisch begrüßten, sind relativ rar, obwohl die Nationalsozialisten von Beginn an eine Aufhebung des Gegensatzes zwischen Natur und Technik propagierten. Gegenüber der Bedeutung des Jahres 1933 hat Frank Uekötter jüngst die These aufgestellt, dass weniger die Machtübernahme als vielmehr der Erlass des Reichsnaturschutzgesetzes eine entscheidende Zäsur für die Haltung der Na-

[36] *Walther Schoenichen*, Naturschutz im Dritten Reich, Berlin 1934, X, 3 f.

[37] Vgl. *Uekötter* (wie Anm. 1), 45 ff.

[38] *Wilhelm Heinrich Riehl*, Die Naturgeschichte des Volkes als Grundlage einer deutschen Sozial-Politik, Bd. 1: Land und Leute, Stuttgart [9]1894, 52. Freilich schrieb der Begründer der Volkskunde im Gegensatz zu den Nationalsozialisten aus der Grundhaltung eines sozialpolitischen Konservativismus über den *„Zusammenhang von Land und Volk als Fundament aller sozialen und politischen Entwicklung"*. Ebd., X.

[39] Vgl. *LA Schleswig* Abt. 301 Nr. 4065, Schreiben des Reichsforstmeisters an den Kieler Oberpräsidenten v. 24.1.1936.

[40] Vgl. *Hubertus Neuschäffer*, Geschichte von Wald und Forst in Schleswig-Holstein und der alte Rendsburger Wald mit dem Forstamt Barlohe, Rendsburg 1986, 48.

turschützer zu den Nationalsozialisten darstellte: „[...] *the distance between the two camps shrank dramatically in 1935, giving way to a strong affection, if not enthusiasm, for the Nazis. [...] Praise for the new law was almost universal, and it pertained to both the law's general intentions and its specific provisions*“[41].

Reichsforstmeister Göring hatte sich 1935 massiv für das Gesetz eingesetzt und die Zuständigkeit für Jagd und Naturschutz an sich gezogen. Erstmals wurden im Reichsnaturschutzgesetz die unterschiedlichen Schutzkategorien, die sich in verschiedenen Provinzen Deutschlands herausgebildet hatten, reichsweit einheitlich kodifiziert. Neben Naturdenkmälern und Naturschutzgebieten konnten nun auch Landschaftsteile mit geringerem Schutzniveau ausgewiesen werden. Zu den ersten der neu ausgewiesenen Flächen in Schleswig-Holstein gehörten mehrere Hünengräber in der Nähe Aukrugs. Als eines der wenigen relativ großen Naturschutzgebiete wurde 1938 die Hahnheide bei Trittau ausgewiesen (1.450 ha). Nach und nach wurden nun zudem viele der älteren Naturschutzgebiete in das Reichsnaturschutzbuch eingetragen[42]. Bei allen naturschutzrelevanten Planungen waren seit 1935 die Naturschutzbehörden zu beteiligen. Unklar ist, ob bei der Ausweisung neuer Naturschutzgebiete häufiger auch ein gänzlich neues scharfes Instrument eingesetzt wurde. Im Gegensatz zu bereits existierenden Landesgesetzen ermöglichte das Reichsnaturschutzgesetz nämlich die entschädigungslose Enteignung von Privatbesitz[43]. Bevor entsprechende Fallstudien vorliegen, kann allerdings davon ausgegangen werden, dass die alleinige Existenz dieses Passus die Verhandlungsbereitschaft zum Verkauf potentieller Naturschutzflächen erhöht haben müsste.

Was aber bedeutete die NS-Zeit unterhalb der rechtlichen Ebene ganz konkret für die Schleswig-Holsteinische Natur? Im Folgenden soll dies kurz für den Bereich der Moore und die Naturschutzgebiete auf der Insel Sylt skizziert werden.

Wie bereits angedeutet, war den Naturschützern der Schutz einiger der letzten verbliebenen Moore ein besonderes Anliegen. Wie in anderen Landesteilen war die Naturlandschaft der Mooroberfläche durch das Torfstechen und umfangreiche Entwässerungsmaßnahmen bedroht[44]. Natürlich konnte man nicht wie im alpinen Raum auf besonders ursprüngliche Hochmoore verweisen. Doch zu Beginn der 1940er Jahre wurden einige Moorgebiete wie das Groß Wittenseer Moor und das Kaltenho-

41 *Uekötter* (wie Anm. 1), 52 f.

42 Das nach dem Führerprinzip gestaltete Verfahren dafür lief so, dass der Leiter der Reichsstelle für Naturschutz bei den Landesbeauftragten Gutachten über infrage kommende Gebiete anforderte, die er an den Reichsforstmeister weiterleitete. Dieser ermächtigte dann die Regierungspräsidenten zum Erlass entsprechender Naturschutzverordnungen. *BA Koblenz* B 245-196.

43 Vgl. *Ludwig Sick*, Das Recht des Naturschutzes. Eine verwaltungsrechtliche Abhandlung unter besonderer Berücksichtigung des preußischen Rechts mit Erörterung der Probleme des Reichsnaturschutzgesetzes, Diss., Bochum 1935, 83 ff.; siehe weiterführend *Edeltraud Klueting*, Die gesetzlichen Regelungen der nationalsozialistischen Reichsregierung für den Tierschutz, den Naturschutz und den Umweltschutz, in: Naturschutz und Nationalsozialismus, hg. v. Joachim Radkau und Frank Uekötter, Frankfurt/Main 2003, 77-105.

44 Vgl. *H.E. Jungjohann*, Über die Regeneration des Kaltenhofer Moores. Ein Beitrag zur Moor- und Vegetationskunde Schleswig-Holsteins, Staatsexamensarbeit Botanisches Institut Univ. Kiel 1953.

fer Moor (1942) bei Dänischenhagen unter Schutz gestellt. Als gegenläufig zu diesen Schutzmaßnahmen erwies sich jedoch die NS-Autarkiepolitik, die die Ausbeutung heimischer Ressourcen substantiell verstärkte. Auch in Schleswig-Holstein scheint sich der Druck auf die Moore erhöht zu haben. So betrieb der Reichsarbeitsdienst gegen den Protest von Naturschützern in den 1930er Jahren vielerorts die Kultivierung so genannten „*Ödlands*", worunter auch viele kleinere ungenutzte Moore fielen. Erst nach Erfolgen in der Eroberung *„neuen Lebensraums"* ordnete Hitler 1941 an, *„daß die Moorkultivierung gestoppt werden sollte, da die Deutschen nun Land genug bekämen"*[45]. Umgehend benachrichtigte der Chef der Reichskanzlei den für die *„Erzeugungsschlacht"* zuständigen Reichsernährungsminister Walther Darré und forderte ihn auf, die noch vorhandenen Moore nicht mehr zu kultivieren[46]. Auf Provinzialebene versuchte der Landesbeauftragte für Naturschutz, Walther Emeis, dem Führerwort Geltung zu verschaffen. Er wurde jedoch von der Landesbauernschaft ausgebremst[47].

Ähnlich ambivalent erging es dem Wald. Zwar wurde im äußerst waldarmen Schleswig-Holstein in den 1930er Jahren das Tempo der Aufforstungen in den Staatsforsten erhöht. Wie Neuschäffer schreibt, wurde allerdings bereits Ende 1936 der Holzeinschlag auf 150 Prozent des normalen Hiebsatzes heraufgesetzt – eine Entwicklung, die im Zeichen der Kriegswirtschaft noch beschleunigt wurde[48]. Ein weiteres Beispiel, das der umwelthistorischen Untersuchung wert wäre, ist die beschleunigte Landgewinnung. In den 1930er Jahren wurden mit dem *„Adolf-Hitler-Koog"* und dem *„Herrmann-Göring-Koog"* große Gebiete an der Nordseeküste gewonnen und später eingedeicht.

Fallbeispiel: Naturschutz auf der Insel Sylt

Während großräumige Bewertungen des schleswig-holsteinischen Naturschutzes im Nationalsozialismus angesichts der Forschungslage zum gegenwärtigen Zeitpunkt noch nicht möglich sind, lassen sich die ambivalenten Wirkungen der NS-Zeit auf die Sylter Inselnatur und die meist zugezogenen Protagonisten, die für ihren Schutz eintraten, aufzeigen.

Unter den nordfriesischen Inseln war Sylt wegen der herausragenden Vogelwelt bereits im Kaiserreich in den Blick der Naturschutzbewegung geraten. Nach dem 1.

45 *Radkau* (wie Anm. 5), 298. – Für die Einschätzung des Leiters der Reichnaturschutzstelle siehe *Hans Klose*, 50 Jahre staatlicher Naturschutz, Gießen 1957, 32, 35.

46 *BA Koblenz* B 245-232, Abschrift des Schreibens von Lammers an Darré v. 25.2.1941.

47 *BA Koblenz* B 245-232, Abschrift des Schreibens von Emeis an Klose v. 6.6.1942. Klose antwortete: *„Die Abweisung von Einsprüchen unter Berufung auf die Willensäußerung des Führers unterliegt natürlich gewissen Bedenken, aber ich möchte glauben, dass die Naturschutzbehörde nicht übertrieben vorsichtig zu sein braucht. Tatsache ist doch, daß der Führererlaß besteht, und zweite Tatsache, daß er wenigstens formell nicht widerrufen wurde."* Schreiben von Klose an Emeis v. 18.6.1942.

48 Vgl. *Neuschäffer* (wie Anm. 40), 48.

Weltkrieg erwarb die Freideutsche Jugend von der Reichsmarine *„für geringes Geld“* ein malerisches Lagergelände in der Nähe des Ortes List. Die Marine hatte nämlich befürchtet, dass, falls Sylt in den Volksabstimmungen 1920 an Dänemark fiele, ihre Anlagen entschädigungslos verloren gingen[49].

Seit dem Ersten Freideutschen Jugendtag im Oktober 1913 auf dem Hohen Meißner wurde *„freies Wandern“* als Weg gesehen, *„ein inniges Verhältnis zu Natur und Volkstum“*[50] auszudrücken. Nach dem 1. Weltkrieg versuchte die Freideutsche Jugend den Jugendlichen, die durch den Krieg und Militärdienst gelitten hatten, eine Heimat zu geben. Als Gründer des Freideutschen Jugendlagers Klappholttal setzte sich der jugendbewegte Arzt Knut Ahlborn nach 1920 auf Sylt für neue Formen der Naturwahrnehmung ein. Neben Maßnahmen der Wohlfahrtsämter diente das Lager der *„wandernden Jugend“* zur Unterkunft. Kritisch beäugt von den Inselbewohnern fanden von 1921 bis 1924 3700 Jugendliche und Erwachsene und über 1900 Kinder im Lager Erholung[51]. Zusammen mit dem Dürerbund und dem Verein Naturschutzpark setzte sich Ahlborn bald dafür ein, mehrere Gebiete auf Sylt unter Naturschutz zu stellen. 1923 wurden nicht nur das gesamte Dünengebiet des so genannten Listlandes und die südlich anschließenden Kampener Dünen bis Kliffende rechtlich geschützt. Auch am Morsumkliff wurde jeder Eingriff in Grund und Boden und jedes Stören und Schädigen der Pflanzen- und Tierwelt untersagt. Insbesondere die Natur am Ellenbogen erholte sich von den Kriegseinwirkungen und entwickelte sich in den nächsten Jahren zum bedeutendsten Vogelschutzgebiet in Schleswig-Holstein[52]. Für die Betreuung dieser Gebiete und weiterer vorgeschichtlicher Kulturdenkmäler gründete man u.a. zusammen mit dem aus der Hygienebewegung bekannten Sanitätsrat Georg Bonne[53] aus Hamburg und dem Provinzialkommissar für Naturschutz, Emeis, den *Verein Naturschutz Sylt*[54].

Obwohl Ahlborn im *„Dritten Reich“* in die NSDAP und als Sturmarzt in die SA eintrat, geriet er wegen seiner freikörperkulturellen Gesundheitspraktiken bald ins Visier der Kreisleitung der NSDAP und der Polizei. Als kolportiert wurde, dass im Rahmen der Sport- und Gesundheitsprogramme am Strand von Klappholttal beide Geschlechter unbekleidet badeten, wurde er vom Kreisleiter in Niebüll vorgeladen. Bald darauf wurde gegen ihn eine Polizeistrafe und wegen *„grober Disziplinlosigkeit*

49 *BA Koblenz* B 245-96, fol. 376, Schreiben von Dr. v. Bahrfeld an die Staatliche Stelle für Naturdenkmalspflege v. 27.7.1921. Zur Gründungsphase des Jugendlagers und einer finanziellen Unterstützung des Preußischen Ministers für Volkswohlfahrt siehe *LA Schleswig* Abt. 309 Nr. 21188.

50 *LA Schleswig* Abt. 605 Nr. 743, *Knud Ahlborn*, Kurze Chronik der Freideutschen Jugendbewegung 1913-1953 zu ihrem 40. Jahrestag, o.O. 1953.

51 Vgl. *LA Schleswig* Abt. 605 Nr. 743.

52 *BA Koblenz* B 245-64, fol. 94.

53 Bonne setzte sich besonders für die Reinhaltung der Elbe ein und ist auch für andere Kapitel der schleswig-holsteinischen Umweltgeschichte wie die Stadthygiene von zentraler Bedeutung.

54 *BA Koblenz* B 245-96, fol. 54, Flugblatt *„Sylts Urnatur in Gefahr!“*. Seit 1921 korrespondierte Ahlborn auch mit Hugo Conwentz, um Mitstreiter gegen die Verlagerung eines Militärlagers in ein Wäldchen bei der Kampener Vogelkoje zu finden. *BA Koblenz* B 245-96, fol. 379, Schreiben von Ahlborn an Conwentz v. 1.7.1921.

und Hintergehung der Kreisleitung" der Ausschluss aus der NSDAP verhängt[55]. Die Zukunft des Nordseelagers war stark gefährdet, weil ein Gauleiter zugesagte Kindergruppenbelegungen zurückzog. Ein späteres Verfahren gegen Ahlborn wegen unbefugten Tragens des Parteiabzeichens wurde vom Gaugericht aus formalen Gründen eingestellt[56].

Nachweislich begriff die Gemeinde Kampen die Machtübernahme der Nationalsozialisten bereits 1933 als Chance, um den Schutz der Nordwestheide an der Grenze zum bestehenden Naturschutzgebiet im Sylter Norden vor Bebauung durchzusetzen. Um die kostspielige Entschädigung an die Nordseebad Kampen GmbH zu sparen, bat der Gemeindevorsteher Franzen im Dezember 1933 Herrmann Göring, damals noch preußischer Ministerpräsident, um Hilfe. Dessen „*innerliche Anteilnahme an der Heimatschönheit unseres Ortes Kampen und seiner eigenartigen Naturumgebung*" bei seinem letzten Sommerbesuch hatte ihn ermutigt, Görings „*mächtige Hilfe in einer für Kampen ungemein bedrohlichen Lage in Anspruch zu nehmen*"[57].

Ernsthafte Konflikte mit den Vereinszielen der Naturschützer verursachte nicht nur der Tourismus, sondern auch die Kriegspolitik der NS-Regierung. Da Sylt als einer der Hauptabflugplätze für Einsätze gegen England genutzt wurde, nahm die Wehrmacht bei ihren umfangreichen Baumaßnahmen auf die Vorschläge der *Reichsstelle für Naturschutz* zum Landschaftsschutz kaum Rücksicht. Nach § 6 des Reichsnaturschutzgesetzes durften durch den Naturschutz „*Flächen, die ausschließlich oder vorwiegend Zwecken der Wehrmacht* [...] *dienen, in ihrer Bedeutung nicht beeinträchtigt werden*". So wurde nicht nur der Fliegerhorst Westerland massiv ausgebaut[58]. Der Leiter der Reichsstelle, Hans Klose, schrieb 1940 an die schleswig-holsteinischen Naturschützer: „*Von meinen Vorschlägen in Bezug auf weitgehenden und großräumigen Landschaftsschutz wird dann wohl nicht viel aufrecht zu halten sein. Gewiss, ein Jammer vom Standpunkt des Natur- und Heimatschutzes aus gesehen, aber wer mag augenblicklich darüber klagen, wo es um so gewaltige andere*

[55] Sofort schrieb Ahlborn an den Reichsführer SS und Chef der Deutschen Polizei, Heinrich Himmler, um sich gegen den Ausschluss zu wehren. Vgl. *BA Koblenz* B 245-96, fol. 159, Schreiben von Ahlborn an den Reichsführer SS und Chef der Deutschen Polizei Heinrich Himmler. Im Zweiten Weltkrieg wurde Ahlborn als Truppenarzt u.a. am Westwall eingesetzt. *BA Koblenz* B 245-96, fol. 100, Schreiben von Klose an Ahlborn v. 8.5.1940.

[56] *LA Schleswig* Abt. 358 Nr. 708, Vermerk des Oberstaatsanwalts v. 17.9.1938. Nach dem Krieg rechtfertigte Ahlborn solche Beitritte zur NSDAP damit, dass viele Jugendbewegte „*in positiver Mitarbeit die Bewegung auf bessere und gute Bahnen*" zu drängen versuchten. „*Als besonders tragisch muß es empfunden werden, daß der idealistische Schwung und die soziale Haltung der in der NS-Bewegung mitarbeitenden Freideutschen dieser Bewegung auch in ihren mehr und mehr um sich greifenden bösartigen und unheilvollen Bestrebungen und Auswirkungen indirekt Hilfsdienste leisteten*". *Ahlborn* (wie Anm. 50), 13.

[57] *LA Schleswig* Abt. 320 Südtondern Nr. 519, Abschrift eines Schreibens von Franzen an Göring v. 5.12.1933.

[58] Zur Geschichte der militärischen Ausbaumaßnahmen bis zum Ende des 2. Weltkrieges siehe *Harald Voigt*, Die Festung Sylt. Geschichte und Entwicklung der Insel Sylt unter militärischem Einfluß 1894-1945, Bredstedt 1992.

Dinge geht“[59]. Bei einer Bereisung der Abteilung Landschaftsgestaltung des Festungsbaustabes wurde festgestellt, dass durch die Baumaßnahmen *„ein großer Teil fruchtbarsten Ackerbodens und besonders ein großer Prozentsatz der natürlichen Heiden in Anspruch genommen*“ wurde. *„Zahlreiche Pflanzungen sind durch Rollfelder ersetzt.*“[60] Auch aus ästhetischen und tierschützerischen Gründen prangerten die Berichterstatter die traditionelle Schafhaltung in den Dünen an, die im strengen Winter 1940/41 zu zahlreichen Todesfällen unter den Schafen geführt hatte, was *„eines Kulturvolkes*“ für *„unwürdig*“[61] angesehen wurde. Das Kerninteresse an der Inselvegetation fußte allerdings auf dem militärischen Erfordernis, die Gefechtsbereitschaft der Flak-Batterien durch Aufforstungen sicherzustellen, weil diese mancherorts stark unter Sandflug litt.[62]

Nach dem verlorenen Krieg wirkten sich in den Augen der Naturschützer die von britischen Sprengkommandos zerstörten Befestigungsanlagen besonders negativ auf das Landschaftsbild aus; denn ihre Trümmer *„verunzier[t]en*“ für mehrere Jahre *„das ganze Gebiet*“ und trugen zu den *„geradezu katastrophalen Zuständen*“ im *„Vogelschutzgebiet Ellenbogen*“[63] bei.

Zudem nahmen unmittelbar nach der deutschen Kapitulation britische Besatzungstruppen den Ellenbogen in Beschlag und planierten einen Teil des Naturschutzgebiets für einen Truppenübungsplatz der *Royal Air Force*. Ahlborn, der bereits vor Kriegsende militärische Eingriffe in die Naturausstattung der Insel kritisiert hatte, versuchte zunächst die Kampener Vogelkoje – eine traditionelle Fangvorrichtung, die im 19. Jahrhundert zum Fang von Seevögeln gedient hatte und 1935 zum Naturschutzgebiet erklärt worden war – vor weiteren Beschädigungen zu schützen. Angesichts der Materialknappheit, des Brennholzbedarfes und des geringen Stellenwerts des Naturschutzes bei den Behörden war dies kein leichtes Unterfangen[64]. Erst Jahre später – nach Wiedererlangung der staatlichen Souveränität – konnten die Naturschützer Hilfe von oben mobilisieren. 1955 setzte sich der deutsche Außenminister auf Betreiben der nunmehr zur Bundesanstalt für Naturschutz und Landschaftspflege aufgewerteten ehemaligen Reichstelle für Naturschutz beim britischen Verbindungskommando in Bonn dafür ein, *„daß britische Soldaten das Naturschutzgebiet auf der Ellenbogenhalbinsel nördlich von List keinesfalls betreten* [*sollten*], *damit*

[59] So die Einschätzung von Klose. *BA Koblenz* B 245-96, fol. 100, Schreiben von Klose an Ahlborn v. 8.5.1940.

[60] *LA Schleswig* Abt. 734.3 Nr. 773, Bericht über die Dienstreise der Abteilung Landschaftsgestaltung (Uffz. Harms und Soldat Loebell) vom 16.-23.6.1941 nach Sylt, fol. 124 ff. hier fol. 127.

[61] Ebd., fol. 125.

[62] Vgl. auch *LA Schleswig* Abt. 734.3 Nr. 773, Schreiben des Marschenbauamts Husum an das Kommando der Marinestation der Nordsee – Oberfestungsbaustab – in Wilhelmshaven v. 25.9.1941, fol. 133 ff.

[63] So die zeitgenössische Bewertung von Walther Emeis, in: *BA Koblenz* B 245-64, fol. 93-95, Bericht zum Stand der Naturschutzarbeit von Walther Emeis in der Niederschrift über den erstmaligen Zusammentritt der Landesstelle für Naturschutz und Landschaftspflege in Schleswig-Holstein am 15.3.1949.

[64] *BA Koblenz* B 245-96, fol. 95, Schreiben von Ahlborn an Klose v. 10.6.1946.

die Vogelbrut nicht gestört werde"[65]. Mit der Einrichtung einer neuen deutschen Inselgarnison bald nach der Wiederbewaffnung drohten neue Naturzerstörungen.

Am Beispiel des Inselnaturschutzes lassen sich nur einige Facetten der wenig untersuchten Themen „*Naturschutz im Krieg*" und „*Naturschutz und Militär*" andeuten. Historisch aufgearbeitet ist vor allem der massive Naturschutzprotest der unmittelbaren Nachkriegszeit gegen die britischen Bombardierungen auf dem Großen Knechtsand im südlichen Wattenmeer westlich von Cuxhaven[66]. Es fehlt eine Untersuchung kleinerer Scharmützel im Bereich des schleswig-holsteinischen Wattenmeers.

Für eine Bilanz der zweiten Schlüsselperiode des deutschen Naturschutzes, der Zeit des Nationalsozialismus, ist es noch zu früh. Die Zahl der schleswig-holsteinischen Gebiete, die auf Grundlage des Reichsnaturschutzgesetzes zu Naturschutzgebieten erklärt wurden, stieg bis Kriegsende zwar von sieben (1933) auf 44[67]. Angesichts der rücksichtslosen Autarkiepolitik und der Zerstörungen durch den Krieg sagt diese Zahl für sich genommen jedoch äußerst wenig über den tatsächlichen Schutz aus[68]. Gerade in strategisch bedeutsamen Gebieten wie der Insel Sylt zeigte es sich, dass Naturschutzgebiete und Windschutzpflanzungen zum Teil sogar als Tarnung in den Ausbau zur Festung mit zahlreichen Flak-Batterien einbezogen wurden[69].

Im Anschluss an die Forschungen von Joachim Wolschke-Bulmahn und Gerd Gröning über die Reichsstelle für Raumordnung wurde wiederholt dargelegt, wie sich die Verantwortlichen des Naturschutzes auf höchster Ebene auch an der Umgestaltung der eroberten Gebiete in Mittel- und Osteuropa beteiligten[70]. Der Menschen verachtende Geist, den Himmlers „*Allgemeine Anordnung über die Gestaltung der Landschaft in den eingegliederten Ostgebieten*" von 1942 atmet, lässt sich meines Erachtens nicht trennen von den NS-Naturschutzkonzeptionen im „*Altreich*". Die Allgemeine Anordnung, die auf die Aktivitäten des „*Landschaftsgestalters*" Heinrich Friedrich Wiepking-Jürgensmann zurückgeht und u.a. auf der Theorie der angeblichen „*kulturellen Überlegenheit des deutschen Volkes*", kombiniert mit einer aggres-

65 *BA Koblenz* B 245-96, fol. 30, Abschrift eines Schreibens des Bundesverteidigungsministeriums an die Landesregierung Schleswig-Holsteins v. 25.6.1955.

66 Vgl. *Engels* (wie Anm. 1), Kap. 4.

67 *BA Koblenz* B 245-64, fol. 93-95, Bericht zum Stand der Naturschutzarbeit von Walther Emeis v. 15.3.1949.

68 Zum Problem der Wertung für Süddeutschland siehe *Richard Hölzl*, Naturschutz in Bayern: Der Landesausschuß für Naturpflege und der Bund Naturschutz zwischen privater und staatlicher Initiative, Regensburg 2005: http://www.opus-bayern.de/uni-regensburg/volltexte/2005/521/.

69 Vgl. *LA Schleswig* Abt. 734.3 Nr. 773, Acta betreffend Dünen-Aufforstungen Correspondenz 1901-1942.

70 Vgl. *Gerd Gröning* und *Joachim Wolschke-Bulmahn*, Die Liebe zur Landschaft. Teil III. Der Drang nach Osten. Zur Entwicklung der Landespflege im Nationalsozialismus und während des Zweiten Weltkrieges in den eingegliederten Ostgebieten, München 1987 und *Klaus Fehn*, „Lebensgemeinschaft von Volk und Raum": Zur nationalsozialistischen Raum- und Landschaftsplanung in den eroberten Ostgebieten, in: Naturschutz und Nationalsozialismus (wie Anm. 1), 207-224. Zum Hintergrund und zur neuesten Einordnung der Allgemeinen Anordnung siehe *Uekötter* (wie Anm. 1), 133 ff.

siven Rassenpolitik, fußt, beschreibt die *„Landschaft in den eingegliederten Ostgebieten* [...] [*als*] *auf weiten Flächen durch das kulturelle Unvermögen fremden Volkstums vernachlässigt, verödet und durch Raubbau verwüstet* [...]. Dem *„germanisch-deutschen Menschen"* sei aber *„der Umgang mit der Natur ein tiefes Lebensbedürfnis. In seiner alten Heimat und in den Gebieten, die er durch seine Volkskraft besiedelt und im Verlauf von Generationen geformt hat, ist die harmonische Gestaltung von Hofstatt und Garten, Siedlung, Feldflur und Landschaft ein Kennzeichen seines Wesens"*. Aus diesen ideologischen Überlegungen wurde die Strategie abgeleitet, nicht nur *„fremdes Volkstum auszuschalten"*, sondern den Osten in einer Weise umzugestalten, dass *„der germanisch-deutsche Mensch sich heimisch fühlt, dort sesshaft wird und bereit ist diese seine neue Heimat zu lieben und zu verteidigen"*[71].

Bereits vor dem Erlass der Allgemeinen Anordnung hatte die Reichnaturschutzstelle Anspruch erhoben, *„zu einer Reichsspitze für die Landschaftsgestaltung ausgestaltet zu werden"*[72]. Sie konnte sich allerdings nur partiell gegenüber dem *„Reichskommissar für die Festigung deutschen Volkstums"*, Himmler, durchsetzen.[73] Zur Umsetzung der Anordnung in den eroberten Gebieten, die auch als *„Grundgesetz"* angesehen wurde, bemühte sich der zuständige Referent in der Reichsstelle für Raumordnung, Erhard Mäding, in der Folgezeit um *„eine enteignungsähnliche gesetzliche Grundlage nach Art des Naturschutzgesetzes"*. Er stellte aber auch bereits taktische Überlegungen an, wie das neue *„Landschaftsrecht"* nach dem Krieg allmählich *„auf das gesamt*[e] *Reich erstreckt"*[74] werden könnte. Die Rechtsabteilung empfahl *„einstweilen"* zu versuchen, *„die Landschaftsgestaltungsmaßnahmen in der Praxis anlaufen zu lassen. Bei 80-90% des Grundeigentums in den eingegliederten Ostgebieten können wir diese Massnahmen auf Grund unseres Rechts zur Verfügung über Grund und Boden (Polenverordnung) oder durch Einflussnahme auf die Ostland-Siedlungsgesellschaften, Domänen-Abteilung und die sonstigen öffentlichen Stellen, die im Osten Grundbesitz bewirtschaften, durchführen.* [...] *Wenn einige wenige volksdeutsche Querköpfe sich widersetzen, so dürfte dadurch das Gesamtbild und die Gesamtarbeit in den eingegliederten Ostgebieten wenig beeinträchtigt werden."*[75]

71 *BA Koblenz* B 245-88, Reichsführer SS/Reichskommissar für die Festigung deutschen Volkstums, Allgemeine Anordnung Nr. 20/VI/42 über die Gestaltung der Landschaft in den eingegliederten Ostgebieten vom 21. XII. 1942.

72 *BA Berlin* R 49-511, Schreiben von Wiepking-Jürgensmann an den Reichsführer SS v. 11.12.1941.

73 *BA Berlin* R 49-512, Vereinbarung zwischen dem Reichskommissar für die Festigung deutschen Volkstums – Stabshauptamt und dem Reichsforstmeister als Oberster Naturschutzbehörde v. 11.5.1942.

74 *BA Berlin* R 49-512, Vermerk Vorg.: Recht der Landschaft v. 17.8.1942. Zur zeitgenössischen programmatischen Einordnung siehe *Erhard Mäding*, Das Recht der Landschaft. Anregungen zur Rechtsgestaltung der Landschaftspflege, ca. 1942, in: ebd. Mädings Landschaftsvorstellungen waren wohl auch von seiner Zeit im *„Jungdeutschen Orden"* geprägt. Vgl. *Joachim Wolschke-Bulmahn*, Die Landschaftsideale der bürgerlichen Jugendbewegung und ihre Bedeutung für Landschaftspflege und Naturschutz, in: Wegmarken: Beiträge zur Geschichte des Naturschutzes; Festschrift für Wolfram Pflug, Essen 2000, 18-40, hier 20.

75 *BA Berlin* R 49-512, Schreiben der Rechtsabteilung an Amt VI z. Hd. v. Herrn Dr. Mäding v. 8.10.1942.

Durch den Kriegsverlauf wurde die Umgestaltung des Ostens zur Heckenlandschaft für den *„germanisch-deutschen Menschen"* und zugleich die Übertragung der Konzepte und Kompetenzen auf Reichsgebiete wie Schleswig-Holstein abgewendet. Für eine Bewertung des Naturschutzes im Nationalsozialismus dürfen die angedeuteten verbrecherischen und rasseideologischen landschaftspflegerischen Konzeptionen keinesfalls ausgeblendet werden.

III. Zwischen Reorganisation des Naturschutzes und umweltpolitischer Wende der frühen 1970er Jahre: Zur Vorgeschichte des Nationalparks Schleswig-Holsteinisches Wattenmeer

In der sowjetischen Besatzungszone setzte die Deutsche Verwaltung für Volksbildung direkt nach dem Krieg einige Naturschutzbücher etwa von Walther Schoenichen auf die *„Liste der auszusondernden Literatur"*[76]. Im Westen firmierte die Reichsstelle für Naturschutz und Landschaftspflege nach 1945 wieder als Zentralstelle. Klose gelang es, seine Institution in den nächsten Jahren im Behördenapparat des Bundes zu verankern. Zugleich beriet er die Obersten Naturschutzbehörden, die nun bei den einzelnen Bundesländern angesiedelt wurden, weiterhin in Rechtsfragen. Eine weitere Aufgabe der Zentralstelle war die *„Wahrnehmung der deutschen Interessen im internationalen Naturschutz"*[77]. Das Reichsnaturschutzgesetz behielt in modifizierter Form nach 1945 zunächst als Bundes- und dann als Landesrecht seine Gültigkeit. Bereits 1946 vermutete Klose, *„dass man dem Gesetz die autoritären Zähne ziehen und durch demokratische Goldzähne ersetzen wird."* Über das neue politische System äußerte er sich zunächst skeptisch: *„Ich muss nach bestimmten Mitteilungen befürchten, dass die unterschiedlichen Parlamente eingeschaltet werden. Das wäre nicht das schlimmste, wenn wir ein Zweiparteiensystem bekämen wie das die Engländer oder Amerikaner haben. Aber bei dem Gedanken, dass 5 bis 17 Parteien mitreden werden, kann man das graue Elend bekommen. Vom Naturschutz wird da nicht viel mehr übrig bleiben. Aber diese vertraulichen Ketzereien stehen ja im Augenblick nicht zur Debatte."*[78]

Auf dem Gebiet Schleswig-Holsteins, das mit der Zergliederung Preußens den Status eines eigenständigen Bundeslandes erhielt, änderte sich abgesehen vom deutlich geringeren programmatischen Einfluss des Naturschutzes personell und strukturell recht wenig. Nach 1945 übernahm das Kieler *Ministerium für Ernährung, Landwirtschaft und Forsten* die Funktion der Obersten Naturschutzbehörde. Durch Be-

[76] Berlin 1946.

[77] *BA Koblenz* B 245-64, fol. 399, Schreiben von Klose an die Oberste Naturschutzbehörde des Landes Schleswig-Holstein v. 26.10.1949. Die Reorganisation des Netzwerkes der Naturschützer ist in der Literatur bereits häufiger beschrieben worden. Auch die alten Kontakte zwischen Berlin bzw. nun Egestorf, Bonn und Flensburg erwiesen sich als tragfähig, zumal der langjährige Landesbeauftragte für Naturschutz, Walther Emeis, bis 1968 im Amt blieb.

[78] *BA Koblenz* B 245-66, fol. 241, Brief von Klose an Benick v. 30.7.1946.

schluss des Alliierten Kontrollrates und den angedeuteten Wechsel der Zuständigkeit des Naturschutzes auf die Landesebene galt das Reichsnaturschutzgesetz mit einigen Änderungen auch im Norden bis 1976 als Landesrecht weiter.

Auch wenn in den 1960er Jahren im Bereich des Wattenmeerschutzes neue Akteure auftraten, erwiesen sich erst die frühen 1970er Jahre als Zeit des beschleunigten Wandels. Die Politisierung des Umweltschutzes durch die sozialliberale Koalition im Bund schlug sich auch auf das umweltpolitische Klima in Schleswig-Holstein nieder. Neben den alten Naturschutzverbänden entstand eine Vielzahl von lokalen Umweltinitiativen.[79]

Wie aber wirkte sich die Umweltwelle auf die gestandenen Naturschützer aus? Rückblickend wurden die mentalen Wirkungen der veränderten Umweltwahrnehmung auf den Naturschutz folgendermaßen beschrieben: *„Die plötzlich aufbrechende Wahrnehmung der viel tiefer greifenden ökologischen, funktionalen Störungen im Wachstumsgefolge rief Anfang der siebziger Jahre das weltweite Umweltbewußtsein auf den Plan, dem umgehend Umweltpolitiken und der programmatische Umweltschutz folgten.* [...] *Der klassische Naturschutz* [*wurde in diese Entwicklung*] *geradezu aufgeschreckt hineingezogen und somit in eine konstruktive Identitätssuche gedrängt“*[80].

Während sich der technische Schutz von Luft und Wasser direkt auf die Schäden konzentrierte, hatte der Naturschutz jahrzehntelang auf das *„vor eben diesen Schäden zu bewahrende Naturerbe“*[81] geblickt. Der Naturschutz wurde von vielen vor allem als kulturelles und ethisches Projekt aufgefasst. Der neue ökologisch orientierte Trend basierte demgegenüber auf der Wahrnehmung der abhängigen Stellung des Menschen in der Biosphäre und dem Gefühl der Bedrohung der Menschheit als Ganzes. Dies veränderte auch den Interpretationsrahmen des Naturschutzes. Trotz all der spektakulären technologischen Höhenflüge begannen sich viele Menschen unter dem Einfluss der Medien zu erinnern, dass die Menschheit Teil der Natur ist und somit auch natürlichen Grenzen unterliegt. Jens Ivo Engels hat in seiner Habilitationsschrift die damit verbundenen Veränderungen in der politischen Praxis der Natur- und Umweltschutzvereine en detail erläutert. Mittels einer differenzierten Analyse des Wandels der *„politischen Verhaltensstile“*[82] der Naturschützer gelingt es ihm, den Wandel der 1970er Jahre genauer zu fassen. Gegenüber den beharrenden Tendenzen in traditionellen Naturschutzverbänden bildete für Engels der 1975 gegründete ökologisch orientierte BUND *„gewissermaßen die Summe der Modernisierungsprozesse im Naturschutz“*[83] – samt zeittypischer bewusster Emotionalisierung der Umweltprobleme und neuen Aktionsformen.

[79] Vgl. *Hünemörder* (wie Anm. 1), 182 ff.

[80] *Kurt Egger*, Humanökologie in Praxis und Theorie, in: Gaia 6 (1997) 2, 146-152, hier 146 f.

[81] Ebd., 147.

[82] Unter *„politischem Verhaltensstil“* versteht Engels nach Martin Dinges *„stabile, aber nicht fixierte Verhaltensweisen von Personen und Gruppen angesichts eines lebensweltlichen Problems“*. *Engels* (wie Anm. 1), 15.

[83] Bis 1980 stieg die Zahl der Mitglieder des BUND auf knapp 55.000. Vgl. ebd., Kap. 7.

Im Verlauf der Auseinandersetzung um die Nordseeverschmutzung blieb übrigens auch England davon nicht unberührt. Als eine Konferenz der europäischen Umweltminister in London stattfand, packten die norddeutschen Naturschützer zusammen mit Greenpeace für jeden Minister einen Präsentkorb mit im Spülsaum gesammeltem Müll. Da sehr vieles davon aus England kam, wurde frei nach Elvis Presley das Motto *„return to sender"* gewählt. Über die Aktion wurde sogar in den *Tagesthemen* berichtet[84].

Zur Untermalung der Politisierungsvorgänge in dieser dritten Schlüsselperiode mit seinen neuen Aktionsformen ist die Vorgeschichte des Nationalparks Schleswig-Holsteinisches Wattenmeer aufschlussreich. Nach den Nationalparken Bayerischer Wald (1970) und Berchtesgaden (1978) wurde mit dem Schleswig-Holsteinischen Wattenmeer 1985 der dritte deutsche Nationalpark eingerichtet. Mit heute über 440.000 Hektar ist der Wattenmeerpark etwa 20 Mal so groß wie der Bayerische Wald. Bereits zu Beginn der 1960er Jahre hatten die Gründungsmitglieder der Schutzstation Wattenmeer um Gerd Oetken erkannt, dass das Netz aus kleineren bestehenden Schutzgebieten vom Verein Jordsand und vom Deutschen Bund für Vogelschutz zwar fortlaufend betreut wurde, aber für das gesamte Wattenmeer *„keine zeitgemäße Fortentwicklung des Naturschutzes"* leisten konnte. So veröffentlichte die Schutzstation bereits ein Jahr nach ihrer Gründung 1963 eine Denkschrift mit dem Titel *„Großreservat Halligmeer"*, in der sie die Idee eines ökologisch begründeten, großräumigen Naturschutzes im Wattenmeer präsentierte[85]. Nach Überzeugungsarbeit bei der Landesregierung und dem einheimischen Landrat wurde 1968 zunächst das *„Naturschutzgebiet Nordfriesische Außensände"* ausgewiesen[86]. Im Jahr zuvor war der Öltanker „Torrey Canyon" vor der britischen Küste gesunken und hatte das Medieninteresse für kurze Zeit auch über Großbritannien hinaus auf die Verschmutzung des küstennahen Meeres gelenkt[87].

Die Naturschützer Wolfgang Erz und Oetken, die beide im Deutschen Jugendbund für Naturbeobachtung für den Vogelschutz gewonnen wurden, versuchten nach diesem Teilerfolg die Idee eines Nationalparks im Nordfriesischen Teil des Wattenmeers weiter zu verbreiten, denn es drohten großräumige Eindeichungsvorhaben. Als Erz von dem neu ernannten Bundesbeauftragten für Naturschutz, Bernhard

84 Vgl. Interview des FÖJ-Jahrgangs 04/05 mit dem Gründer der Schutzstation Wattenmeer Gerd Oetken vom 14.6.2005, Manuskript [wird noch publiziert].

85 Auf Basis umfangreicher Vorarbeiten von Uwe Dulz schlug die Vereinigung vor, dass sich das Gebiet über die Nordfriesischen Außensände und die angrenzenden Wattgebiete samt der Halligen Norderoog und Süderoog erstrecken sollte. Zwei Jahre später erweiterte sie ihren Vorschlag für das „Naturreservat Halligmeer" um die Wattgebiete bei der Insel Pellworm und sämtliche nordfriesische Halligen. Die Naturschutzgesellschaft Schutzstation Wattenmeer e.V. gehörte zu den langfristig erfolgreichsten Neugründungen der 1960er Jahre. Sie geht zurück auf ein Netzwerk von jüngeren naturverbundenen Menschen, die sich vor allem aus gemeinsamen Fahrten und Kongressen des 1950 gegründeten Deutschen Jugendbundes für Naturbeobachtung (DJN) kannten. Vgl. *Gert Oetken*, Stationen eines gemeinsamen Weges zum Nationalpark im Schleswig-Holsteinischen Wattenmeer, in: Natur und Landschaft 72 (1997) 1, 45-47.

86 Vgl. ebd.

87 Vgl. *Hünemörder* (wie Anm. 1), 119, 151.

Grzimek, als Mitarbeiter eingestellt wurde, forcierte er die Bemühungen zur Intensivierung der Nationalparkdiskussion. Zunächst plante er 1971 eine vorbereitende Bereisung eines Expertenteams mit Grzimek an der Spitze. Ein Jahr später organisierte er den Deutschen Naturschutztag in Husum, der unter dem programmatischen Motto *„Schafft Deutschlands zweiten Nationalpark!"* stand[88]. Zur Vorbereitung und Beschleunigung der Ausweisung präsentierte Erz auf dem Naturschutztag sein Buch *„Nationalpark Wattenmeer"*. Doch anstelle einer zügigen Realisierung geriet die Initiative in den nächsten Jahren *„erheblich in parteipolitisches Gerangel"*, bis Ministerpräsident Gerhard Stoltenberg das Nationalparkthema in der Rezession Mitte der 1970er Jahre vorerst beerdigte[89].

In der zweiten umweltpolitischen Welle Anfang der 1980er Jahre wurde die Nationalparkidee politisch wiederbelebt. Laut Erinnerung von Gert Oetken griff Landwirtschaftsminister Flessner das Thema wieder auf, nachdem er nach einem Festvortrag von Erz zum Anlass des 20-jährigen Bestehens der Schutzstation Wattenmeer mit diesem ein längeres Gespräch geführt hatte. Im Juli 1985 verabschiedete der Landtag das Nationalparkgesetz. Mit Inkrafttreten des Gesetzes im Oktober 1985 wurden etwa 285.000 Hektar des Wattenmeeres als Nationalpark ausgewiesen. Die ausgedehnten Wattflächen, Außensände und kleinen Halligen bildeten zusammen mit den Vorländereien des Festlandes und den Inseln Föhr, Pellworm und Nordstrand den größten Nationalpark in Mitteleuropa. Freilich gingen in das Gesamtgebiet zahlreiche Flächen ein, die z.T. bereits seit den 1930er Jahren per Verordnung geschützt waren. Zu den frühen Schutzgebieten gehörten die Hamburger Hallig (1930), die Insel Trischen (1932) sowie Teile der Halligen Norderoog (1939) und Südfall (1959). Per Kreisverordnung wurden ab 1977 Landschaftsteile im Dithmarscher Wattenmeer geschützt. Zu Beginn des Jahres 1983 standen damit etwa 70% des schleswig-holsteinischen Wattenmeeres unter Schutz.

Die konkreten Planungen zur Ausweisung eines Wattenmeernationalparks begannen mit einem viel diskutierten Referentenentwurf des Landwirtschaftsministeriums im Jahr 1983. Die Eile der weiteren Gesetzesarbeiten für die Ausweisung des Wattenmeergebietes von der Elbe bis zur deutsch-dänischen Grenze nördlich von Sylt erklärt sich aus dem Wettstreit mit den Planungen auf niedersächsischer Seite[90].

International arbeiten die Anrainerstaaten beim Schutz des Wattenmeers seit 1978 zusammen. Die verantwortlichen Minister Dänemarks, der Bundesrepublik und der Niederlande treffen sich seitdem in regelmäßigen Abständen zu so genannten *„Trilateralen Regierungskonferenzen"*, um die jeweiligen Maßnahmen im Interesse eines umfassenden Schutzes des Wattenmeers zu koordinieren. Als Grundlage der Zusammenarbeit wurde 1982 in Kopenhagen eine *„Gemeinsame Erklärung zum Schutz des Wattenmeers"* unterzeichnet. 1987 eröffneten die beteiligten Staaten in Wilhelmshaven ein *„Gemeinsames Wattenmeersekretariat"*[91].

[88] Vgl. *Oetken* (wie Anm. 85), 46.
[89] Ebd.
[90] Vgl. Anzahl und Flächengröße der Naturschutzgebiete in der Bundesrepublik Deutschland, in: Natur und Landschaft 58 (1983) 12, 462; *Oetken* (wie Anm. 85), 47.

IV. Ausblick: Naturschutz und die Gesetzgebung der Europäischen Union

Im Gegensatz zum britischen Naturschutz wirkte sich in Deutschland der mehrmalige politische Systemwechsel stark auf das Auftreten der Akteure im Naturschutz aus. Phasen der Politisierung folgten Phasen der politischen Abstinenz. Seit den späten 1960er Jahren begannen sich die Entwicklungen im Natur- und Landschaftsschutz in England und Deutschland allerdings zu synchronisieren und eine neue Umweltbewegung entstand[92]. Insbesondere internationale Institutionen übten eine vereinheitlichende Wirkung für die vorherrschenden Metaphoriken und Schutzstrategien aus[93].

Die EU, beziehungsweise die EG, besetzte das politische Feld des Natur- und Umweltschutzes relativ spät gegenüber internationalen Organisationen wie der NATO, des Europarates oder der OECD[94]. Dies lag auch an ihrer historischen Ausrichtung auf Fragen der Ressourcenverteilung und der landwirtschaftlichen Produktion im gemeinsamen Wirtschaftsraum. Aufgrund ihrer supranationalen gesetzgeberischen Werkzeuge erwies sich die EG allerdings seit den 1980er Jahren auch in Deutschland als äußerst wichtig für die Weiterentwicklung des Naturschutzes. Plastisch lässt sich dies an den Vorgängen um die so genannte *„Flora-Fauna-Habitat-Richtlinie"* andeuten. Die FFH-Richtlinie ist zusammen mit der älteren Vogelschutzrichtlinie das rechtliche Kernstück zur Schaffung des europaweiten Biotopsystems *Natura 2000*. Nach langen Diskussionen wurde sie von den Regierungschefs Anfang der 1990er Jahre unterschrieben. Da der Kommissionsentwurf der FFH-Richtlinie im Interesse eines kohärenten Systems erstmals mit quantitativen Zielvorgaben den Nutzungsverzicht in zahlreichen natürlichen und naturnahen Lebensräumen einforderte, entstand Ende der 1980er Jahre *„erhebliche Unruhe"*[95] bei den Mitgliedstaaten der Europäischen Gemeinschaft. Im so genannten Sunderer Kreis setzten sich die deutschen Naturschutzverbände frühzeitig mit der FFH-Richtlinie auseinander. In der nach dem Tagungsort benannten *„Erklärung von Sunder"* konstatierten sie:

[91] Gemäß Verwaltungsabkommen hat das Common Wadden Sea Secretariat (CWSS) folgende Aufgaben: *„Hilfeleistung bei den trilateralen Regierungskonferenzen und sonstigen Konsultationen und Sitzungen; Sammlung und Verbreitung von Informationen über Tätigkeiten, die mit erheblichen Auswirkungen auf die natürliche Umwelt im Wattenmeer verbunden sind oder sein können sowie Unterbreitung von Vorschlägen für geeignete Maßnahmen; Erarbeiten von Vorschlägen für eine engere wissenschaftliche Zusammenarbeit"*. In: Natur und Landschaft 64 (1989) 7/8, 365.

[92] Zum deutsch-britischen Vergleich siehe etwa *Karl Ditt*, Naturschutz und Tourismus in England und in der Bundesrepublik Deutschland 1949-1980, in: AFS 43 (2003), 241-274, hier 273 und *ders.*, Nature Conservation in England and Germany 1900-70: Forerunner of Environmental Protection?, in: Contempory European History 5 (1996) 1, 1-28.

[93] *Kai F. Hünemörder*, Vom Expertennetzwerk zur Umweltpolitik. Frühe Umweltkonferenzen und die Ausbreitung der öffentlichen Aufmerksamkeit für Umweltfragen in Europa (1959-1972), 275-296.

[94] Vgl. *Hünemörder* (wie Anm. 1), 126 ff.

[95] Vgl. etwa *Hans-Joachim Mader*, Die Isolation von Tier- und Pflanzenpopulationen als Aspekt einer europäischen Naturschutzstrategie, in: Natur und Landschaft 65 (1990) 1, 9-12, hier 11.

„Diese Richtlinie ist vor allem ein unverzichtbarer Beitrag zur Ausgestaltung der Umweltpolitik unter den Aspekten des zu verwirklichenden gemeinsamen Marktes im Sinne einer notwendigen Symmetrie zwischen wirtschaftlicher und technologischer sowie sozialer, kultureller und insbesondere auch ökologischer Entwicklung"[96].

Viele der Naturschutzverbände, etwa die Deutsche Sektion des Internationalen Rates für Vogelschutz, versuchten in den nächsten Jahren den Kommissionsentwurf inhaltlich unbeschnitten über die Hürden der parlamentarischen Beratung zu bringen[97]. Inzwischen wurden die Erfordernisse der Richtlinie im Bundesnaturschutzgesetz und im novellierten Landesnaturschutzgesetz rechtlich umgesetzt.

Als in den 1990er Jahren die Frist zur Anmeldung wertvoller ortstypischer Naturflächen und Artenvorkommen in der atlantisch beeinflussten Region abzulaufen drohte, entwickelte sich die Sammlung und Übermittlung von Gebieten nach Brüssel zum Politikum. Auch bei den Besitzern von Grünland an der nordfriesischen Westküste brach ein Sturm der Entrüstung gegen das sogenannte *„Verschlechterungsverbot"* los. Die Landwirte sahen sich in ihrem Planungshorizont beschnitten und sprachen von *„kalter Enteignung"*. Statt auf Eigentumstitel durfte nämlich ausschließlich auf die naturschutzfachliche Qualität der zu meldenden Flächen geschaut werden. Aufgrund der Konflikte konnte sich der letzte (grüne) Umweltminister 2002 nur noch mit Personenschutz an die Nordseeküste wagen.

Nach Jahrzehnten des Primats der föderalen Gestaltungskompetenz hat sich die EU-Gesetzgebung und die Arbeit des EU-Gerichtshofes zum wichtigen *„driving factor"* im Natur- und Umweltschutz entwickelt. Offen ist die hochpolitische Frage, über welche Mechanismen die Spannungen abgebaut werden können, die sich mit der größeren Entfernung des Ortes der Rechtssetzung von der praktischen Umsetzung des Naturschutzes verstärkt haben.

96 Vgl. DBV – Abgeschlossene und laufende Aktivitäten des DBV – 1989, in: Natur und Landschaft 65 (1990) 5, 257-258.

97 Vgl. DS/IRV – Arbeitsschwerpunkte des DS/IRV 1988, in: Natur und Landschaft (1988) 5, 226.

David G.C. Allan

From the Penny Post to the Eden Project: the Albert Medallists of the Royal Society of Arts 1864-2003

Though not originally a part of the [Royal] Society of Arts' extensive contribution both institutionally and nationally to the commemoration of the Prince Consort, the annual award of the Albert Medal for "*Distinguished Merit in promoting Arts, Manufactures and Commerce*" has become a means whereby the memory of the Prince's Presidency of the institution (1843-1861) is perpetuated and the civilizing consequences of advances in science and the arts, in which he was so profound a believer, are reflected in the achievements of an individual of genius.

The title of the paper refers to the first award, made in 1864 to Sir Rowland Hill for his postal reform, and to the recognition of Tim Smit's work for ecology in 2003. The former provides a direct link to the Prince's circle of devoted public servants and the latter is of special interest to his great great grandson, the Patron of this Society and the President of the RSA, whose concern for the environment is reflected in the general theme of the 2005 conference.

The awarding of medals in addition to money prizes, or "premiums" as they were called, dated from the earliest years of the [R]SA. The founder, William Shipley, in his 1753 "Scheme" suggested that "*In particular cases, as for very curious and valuable inventions or improvements etc, gold medals may be given*" and three years later the Society adopted Henry Baker's plan for awarding medals of gold, silver and copper "*according to the nature and consequences of the Improvement or Discovery*", which was still in force during Prince Albert's presidency. These medals, like those of kindred institutions came to be regarded by men of science and industry as equivalent, if not superior to the honours bestowed by the Crown on an ever growing number of army officers and civil servants. Prince Albert himself had been concerned by this trend, and had discussed the possible institution of an Order of Merit, though he no doubt believed that state and society distinctions could exist side by side[1].

1 *William Shipley,* A Scheme for putting the Proposals into Execution, 1753; *Henry Baker,* Ms memorandum on medals 1756, RSA Archives; *Sir Henry Trueman Wood,* The History of the Royal Society of Arts, London 1913, 315; *David Guy Charles Allan,* William Shipley, Founder of the Royal Society of Arts, London 1979, 45. For the relationship of Society and State medals see *Lyon Playfair,* Memoirs, ed. W. Reid, London 1900, 156, 164.

By 1863, when the Albert Medal was instituted the "*premium*" system was largely in abeyance and the Society of Arts had come to rely on exhibitions, lectures, examinations and the publication of reports as the means by which it could continue its work. In these circumstances awarding the Society's general medals became less central to its activities and, although the delivery of a lecture of a high quality might earn a silver medal, the attainment of a certain level in an examination one in bronze, those struck in gold were only offered as a part of endowed prizes for attempts to solve certain technical and social problems and were subject to the detailed scrutiny of specialist committees, who more than once failed to find a candidate they considered worthy of a reward. After 1914 offers of endowed medals in gold appear to have ceased and, apart from the Albert Medal which was in a unique category only Society medals in silver and bronze were given.

From 1863 to 1900 the Society's general medal bore the head of the President, Albert Edward Prince of Wales, on the obverse and on the reverse a scroll containing the recipient's name. In 1900 a new head of the Prince was engraved and this continued in use when the Prince ascended the throne, the word "Patron" being substituted for "President". Later monarchs allowed this practice to continue until 1948 when her present Majesty, as Princess Elizabeth, became President and allowed the image of her head to be engraved on the medal. Prince Philip, who succeeded her as President in 1952 followed the same course. The later general medals tended to be more boldly incised and simpler in design than those in use at the beginning of the period[2].

Throughout these years the Society's highest award remained the Albert Medal, struck always in gold and repeating the design which Queen Victoria had herself approved, that is to say on the obverse a head of the Prince Consort as he appeared at the end of his life with the inscription "*Albert, Prince Consort. President of the Society 1842-61*" and on the reverse a group of three neo-classical figures symbolising arts, manufactures and commerce designed by Leonard Wyon surrounded by the name of the Society with the date of the medal's institution MDCCCLXIII given below (*Ill. 4*), the recipient's name being engraved around the edge of the medal.

Except for four years (1867, 1893, 1923 and 1946) when the award was shared by two candidates, the Albert Medal has been given annually to an individual since 1864. Thus up to and including 2003, one hundred and forty four persons have been given recognition for their "*Distinguished Merit in promoting Arts, Manufactures and Commerce*". Considering this total as two periods in which there were 52 awards and one in which there were 49, we arrive at a reasonably satisfactorily chronological sequence: Period One 1864 to 1913, taking us to the "High Noon" of Empire and the eve of the first Great War, Period Two 1914 to 1963, covering both the great conflicts and the world economic depression which separated them, and Period Three 1964 to 2003 in which the Society's home country declined from its position of world leadership but the institution itself increased in size and financial strength.

2 *Trueman Wood* (note 1), 320; *Christopher Eimer*, The Society's President's medals: RSA Jnl 143 (Jan 1995), 69-74. A unique case was the striking of the Society's medal in gold in 1973 for Edgar Lawley.

Ill. 4 The Albert Medal: obverse and reverse (Photograph of the medal awarded to Sir Rowland Hill in 1864 and presented to the RSA by Akio Morita, Albert Medallist 1983: photo RSA).

The first period 1864-1913

Having settled the purpose of the medal early in 1863 and having heard that the Queen had approved its design at the end of the year, the Council of the Society proceeded to select Sir Rowland Hill as the first recipient. This took place on 13 April 1864 when a committee chaired by Sir Joseph Paxton and attended by Henry Cole met and chose Hill's name. Both Cole and the Society as a whole had campaigned for postal reform and Hill, like Cole and Paxton, had a great admiration for the late Prince[3].

An announcement in the Society's *Journal* for 29th April included the text of a letter of thanks and appreciation from Hill:

"Hampstead, N.W. April 22 1864

Sir, I have the honour to acknowledge the receipt of your letter of Yesterday's date, informing me that the Council of the Society of Arts have awarded to me the 'Albert Medal' in recognition of my services in connexion with Postal Reform.

I beg you will do me the favour to express to the Council my grateful acknowledgements of the honour which they have been pleased to confer upon me – an honour

[3] RSA, Council Minutes 4, 13 February, 2 December 1863, 13 April 1864; *Henry Cole*, Ms Diary 13 April 1864; Jnl. SA 11 (1863), 549, 12 (1864) 263; *Elizabeth Bonython* and *Anthony Burton*, The Great Exhibitor, the life and work of Henry Cole, London 2003, Chapter 4; *Trueman Wood* (note 1), 396, 471.

which, whether with reference to the high character of the Society in whose name the award is made, or of the lamented Prince in whose memory the medal has been established, will ever be held by myself and my family in the highest estimation.

I have the honour to be, Sir

Your most obedient servant

ROWLAND HILL"[4].

If as is most likely, this first award resulted from the influence of Cole, the part played by him in regard to the second is well documented and notorious. It was decided that every Vice-President and member of Council of the Society should be invited to submit nominations and, at a meeting held on 29 March 1865, seven names were given preliminary consideration, amongst which were the Emperor of the French Napoleon III, the famous scientist, Michael Faraday, the free trade hero, Richard Cobden, the artist, Daniel Maclise, and the arbiter of taste, John Ruskin. Before a choice could be made by the Council, Cole pre-empted the matter in the course of a visit to Paris when he told Lord Cowley, the British Ambassador, of the Society's desire to award the medal to the Emperor in recognition of "*his services to Arts, Manufactures and Commerce by the abolition of passports*". The result has been described by a modern scholar as "*a rumpus*", involving repeated behind the scenes discussions and manoeuvrings in which Cole's precipitate behaviour caused acute embarrassment for the late Prince's eldest son and namesake who had only recently been elected President of the Society, and who privately expressed the view that to make such an award to the Emperor "*would be a straining of the conditions upon which* [*the*] *medal is granted*". He was, of course, more diplomatic in his dealings with the Council and he commanded his Private Secretary to state: "*His Royal Highness is of opinion that it would be superfluous in him to express any opinion on the subject, since upon enquiry he finds that the Emperor* [...] *has already been communicated with through Lord Cowley at the instance of Mr Cole; so that in point of fact his sanction is now asked to a proposal which may be said to have been already acted upon, and from which* [...] *it would be exceedingly difficult now to recede. At the same time, I am desired to add that should the Council decide upon making the award* [...] *His Royal Highness will, as President, communicate the resolution of the Society to His Majesty*"[5].

Some members of Council still had reservations; the abolition of passports was considered "*an act of state*" and as such to be outside the terms of reference of the Medal. Many doubted if they had the right to confer the medal on a "*Foreign Sovereign*"[6]. Yet, in the end, Cole prevailed and an announcement was published on 30th

[4] JSA 12 (1864), 263.

[5] RSA, Council Minutes 6 May 1865, quoted in *John Sheridan Skidmore*, The Award of the Albert Medal to Napoleon III: Jnl. SA 137 (1989), 652-657.

[6] *Ibid.*, 656.

June stating that the Medal had been awarded to "*His Imperial Majesty the Emperor of the French for distinguished merit in promoting in many ways, by his personal exertions, the international progress of Arts, Manufactures and Commerce, the proofs of which are afforded by his judicious patronage of Art, his enlightened commercial policy, and especially by the abolition of passports in favour of British subjects. His Royal Highness the President has himself communicated this decision to the Emperor, by whom it has been most graciously received*"[7].

Michael Faraday, who had been among the choices in 1865, and who, like Rowland Hill, had a profound respect for the late Prince Consort, was the first of thirty-two eminent scientists to receive the medal in this period. He was followed three years later by the first Foreign scientist, the Hessian Justus von Liebig, whose application of chemical science to agriculture was widely respected in Great Britain, the French Empire, and the German states. In the fateful year 1870, and before the downfall of Napoleon III had occurred, Ferdinand de Lesseps was honoured and, the year after, the ex-Emperor's erstwhile champion, Henry Cole, received the medal from the hands of the Royal President[8].

Sir William Siemens, whom we may think of as Anglo German, was next and then came Michel Chevalier, an erstwhile supporter of Louis Napoleon and still serving the economic advance of France. Our own Astronomer Royal, Sir George Airy, was followed by Jean Baptiste Dumas, an eminent French scientist, and the British engineer, Lord Armstrong, and after two more British scientists, the great Professor Hofmann, sometime protégé of the late Prince Consort, learnt of his honour in a handwritten letter from the Royal President. Louis Pasteur was honoured in 1882 and in 1883 the second of the Botanic "dynasty" of the Hookers received the award. By this date numerous foreign candidates were being considered. Pasteur himself had first been proposed in 1876, following Henry Cole's suggestion that the Foreign Office should ask various embassies and legations abroad to consult about suitable persons. The notable failures include Frederick Krupp, "*Herr*" Semper and (of local interest) Adolphe Besser, the cartographer at Gotha[9].

"*The distinguished American engineer*" Captain James Buchanan Eads, who got the medal in 1884, might have been preceded by E.T. Bigelow, Alven Clark, Cyrus Field, Lewis Rutherford and other recommended United States citizens. Foreign recipients had their medals sent to them through diplomatic channels, but native winners, such as Henry Cole, who noted the event in his diary, were received by the Prince of Wales at Marlborough House. In 1885, however, when the medal had been voted to Sir Henry Doulton, the Prince had what Edmund Gosse called "*one of those*

7 JSA XII (1865), 532.

8 The medal was presented to Faraday by the Chairman William Hawes at Hampton Court House on 16 June 1866. (Information supplied by the Royal Institution) Henry Cole wrote in his diary, "*To see the Pr. Of Wales the first time since his illness. He gave me the Albert Medal with a little speech*", Victoria and Albert Museum, Cole Diary, 6 March 1872.

9 Lists of candidates are given in RSA, Council and Committee Minutes PR.GE/101/10/33-4, Albert Medal Papers. For those chosen, see below, appendix.

luminously tactful inspirations which have done so much to make him popular". He decided to visit Doulton at his Lambeth factory and present the medal before an audience of work people, resulting in a "*seventh heaven of patriotic satisfaction*"[10].

The next year, another man of business received the medal. It was Lord Masham who had been proposed as long ago as 1875 when Henry Cole had been unsuccessful in advocating Sir Titus Salt and the controversial name of Mr Gladstone was rejected. In 1887 four persons were selected as worthy: Alexander Graham Bell, the American inventor of the telephone; Hermann Louis Helmholtz, the German physiologist and physicist; David Hughes, inventor of the printing telegraph and microphone; and John Percy, the eminent metallurgist. All would eventually succeed. But the special occasion of the Queen's Golden Jubilee year precluded their immediate nomination.

Among the proposals made for the Society to commemorate the Golden Jubilee was that it should institute an annual award for women to be called "The Victoria Medal", to be given in the same manner as the Albert Medal with the possible citation being for the woman "*who has done most for the welfare of mankind*". After six months of consultation in which alternative means of commemoration, such as a further statue of the late Prince and a medal for all the members were considered, the idea of a Victoria medal was dropped and it was agreed instead to present the Albert Medal to the Queen. It was a satisfactory compromise, pleasing both the feminists in the Society, the medal enthusiasts, and those anxious to invoke Prince Albert's memory. *The Times* noted that, since the medal had once been given to the late Emperor Napoleon III, there was a precedent for its presentation to a reigning sovereign. Her Majesty was told of the Council's proposal in a memorandum written for her by Sir Henry Ponsonby on 31 May: "*The Society of Arts instituted the Albert Medal as a Memorial of the Prince. It is given yearly to the person who has done most in promoting Arts, Manufactures and Commerce. The Council this year have voted unanimously that it should be offered to Your Majesty as having done more for 50 years than anyone else to promote these matters. The Prince of Wales as President asks if The Queen will graciously accept the medal. If so, it might be given by the Prince of Wales and a Deputation to The Queen at any time this year*".

With the hallowed name so evident in the opening paragraph to her quickly scanning eyes, the Queen wrote her message of approval: "*Shall be very much gratified to receive the medal V.R.I.*"

Having learnt that the Queen would accept the Medal on the 6th June, the Council also settled the texts of a citation and of a congratulatory address. The Medal would be awarded to the Queen "*in commemoration of the progress of Arts, Manufactures and Commerce throughout the Empire during the fifty years of her reign*". She was worthy of it because, as *The Times* put it, she "*may be held to be in this case the personal embodiment of the nation and to represent the aggregate of its work*". The Address spelt out in traditional form the respect which the Society, along with the na-

10 *Edmund Gosse*, Sir Henry Doulton the man of business as a man of imagination, London 1970, 125. Doulton produced a commemorative stoneware vase impressed with a copy of the medal.

tion, felt for the Throne and its occupant. The Prince of Wales arranged for the Council to attend her Majesty at Buckingham Palace on the afternoon of 8 March 1888. Full of anxiety about the news of the German Emperor's rapidly declining health and its implications for the future of her "*beloved Fritz*", the Queen had set out from Windsor to Buckingham Palace at 11 in the morning. "*Saw Lord Salisbury at once on arriving there*", she noted in her *Journal*, and heard about "*the double game Bismarck was playing*". However, "*after luncheon*", turning aside from these family and state worries, she proceeded to the Picture Gallery, where she "*received a gold medal and an Address from the Royal [sic] Society of Arts, Bertie read the address, and I the answer, after which he presented to me the Vice-Presidents and Members of Council*". As we would expect, she accepted "*with pleasure the medal which bears the name of my beloved husband*"[11].

This acceptance established a precedent for numerous similar awards to members of the British Royal Family in this and the succeeding period. It would also be used to justify the continued use of the name "The Albert Medal" when the Home Office unofficially raised the matter of a possible confusion with the *Albert Medal for Gallantry* in 1907[12].

It may seem surprising that no more Foreign Princes were offered the Society's medal and indeed in 1891 the reigning Duke, Ernest II, of Coburg and in 1896 the Austrian Archduke Rainer, allowed their names to go forward. Although their candidatures were unsuccessful, they would have qualified for a special class of Honorary Royal Member into which the German Emperor, the Tsar of Russia and seven other European monarchs were elected early in the next century[13].

Another notable candidate for the Albert Medal in the 1890s was a man who thought little of the niceties of rank. This was William Morris, who had he not died in 1896 would certainly have received it. The Society's Secretary, Sir Henry Trueman Wood recalled that "*he was looked upon* [by the Council] *as a comparatively young man with many years of life before him, and on several occasions his claims were postponed*"[14].

The imperial concerns of the Society which the Prince of Wales thought so worthy were reflected in the awards to pro-consuls, such as the Earl of Cromer (1907) and Lord Strathcona (1912). Feminists and internationalists might have been consoled by the medal granted to Madame Curie (1910) and those who felt that the arts were neglected, when it was given to Walter Crane (1904). Trueman Wood, commenting on the latter point noted that "*The List of Albert medallists, unfortunately, contained*

[11] Royal Archives Windsor Castle RAF 46/48 and Queen Victoria's Journal (8 March 1888) and *The Times* (7 June 1887) quoted *David Guy Charles Allan*, Statues or Medals? The Society of Arts Jubilee Commemoration: Jnl. RSA 135 (1987), 574-578.

[12] See Board of Trade Albert Medal Register 1866-1913, National Archives, Kew; RSA, Council Minutes 2nd October, 4th November 1907.

[13] *John Sheridan Skidmore*, Some Royal Members of the Society 1770-1915, Jnl. RSA 123 (1974), 95-99.

[14] Jnl. SA (27 May 1898), 46, 630.

very few workers in art, and that was due to the fact that it was not sufficient for a man to have done good work, but his work must have exercised great influence"[15]. Such "*great influence*" could sometimes contribute to future catastrophe as was shown in the work of Sir Frederick Abel, Chemist to the War Department, medallist in 1891, and of Sir Andrew Noble, expert on explosives, who gained the award in 1909.

The second period 1914-1963

Apart from the awards to Queen Mary in 1938 and to her present Majesty in 1958, there were no female Albert Medallists in the list until the very end of the second period. This was in spite of the fact that the Council now invited Fellows to suggest the names of "*men and women*" who had achieved "*distinguished merit in promoting Arts, Manufactures and Commerce*". The Arts, which had received scant recognition in the first period, were now honoured in the persons of the architects Aston Webb (1927) and Giles Gilbert Scott (1949), the sculptor William Reid Dick (1948) and the artists Frank Brangwyn (1932) and William Llewellyn (1933) and the designer John A. Milne (1940). Yet the Society generally eschewed the fine arts, qualifying the awards to Brangwyn and Llewellyn with references to commerce and industry. Milne's citation gave respectability to the newly fashionable term "Industrial Art". The award to Ralph Vaughan Williams "*for his eminent services to music*" (1955) is said to have given some surprise to the President, though there were nineteenth century precedents for the Society to concern itself with the training of musicians.

"*Manufactures and Commerce*" were represented by the award to the businessman and car manufacturer Lord Nuffield (1937) and to such eminent British scientists and technologists as Oliver Lodge (1919), Ambrose Fleming (1921), Dugald Clerk (1922), Lord Rutherford (1928), Alexander Fleming and Howard Florey (1946) and Frank Whittle (1952) and Frederick Handley Page (1960). Foreigners such as Marconi (1914), Metchnikoff (1916), Orville Wright (1917) and Sabatier (1926) were also represented, but they all came from the Allied nations of the first Great War, when an extreme xenophobia, so alien to the original ideals of the Society, had caused the expulsion in 1915 from "*the list of Honorary Fellows of all persons of German and Austro-Hungarian nationality*". In the course of his war mission to the USA in 1917 Lord Northcliffe, as a vice-president, presented the Albert Medal to Orville Wright. In the summer of 1941, before the USA became involved in the second great conflict the Society chose President Franklin Roosevelt as the Albert Medallist for that year "*in recognition of his pre-eminent services to humanity as the fearless and resolute champion of the ideals of freedom and individual liberty*". The medal was transmitted to the President through diplomatic channels but when, in 1945,

[15] *Ibid.*

Winston Churchill was chosen, the medal was presented to him in a moving ceremony held in the bomb-damaged library of the Society's House[16].

In the inter-war years, the medal had been given at Clarence House by H.R.H. Prince Arthur, the Duke of Connaught, President of the Society from 1911 to 1942, and sometime Duke of Saxony and Prince of Saxe-Coburg Gotha. Kenneth Luckhurst, who had first attended this ceremony in 1937, was impressed *by "the simple dignity with which the ceremony was performed"*, and by the Duke's invariable reference to the fact that the medal had been *"instituted in memory of my beloved father"*[17]. After an interregnum of three non-royal presidents the present Queen, then Princess Elizabeth, accepted the Presidency and was followed in 1952 by the Duke of Edinburgh. Thus, at the beginning of our third period, the medal would be presented at a ceremony held at Buckingham Palace and then, in consequence of the Duke's ever increasing involvement in the Society, presentations would again be held in the Society's House.

The third period 1964-2003

This period began with the grant of the medal to Dame Ninette de Valois in 1964 *"for her services to the art of ballet"*. She was the first non-royal woman to be honoured since Madam Curie. Six more female awards, three of which were royal, followed in 1974, 1980, 1991, 1998, 2000 and 2001, the last two being to the Princess Royal and to President Mary Robinson of Ireland respectively. The patronage and performance of music was recognised in the awards to Sir Robert Mayer (1979), Yehudi Menuhin (1981), Sir Claus Moser (1996) and Sir Simon Rattle (1997). If Simon Rattle may be accounted among the youngest Albert medallist to date, Sir Robert Mayer, who was more than twice his age and a centenarian, must be accounted the oldest. He was a native of Mannheim where he studied in the Conservatoire, and had contact with Brahms. To escape Prussian militarism he came to England where success in business allowed him to take a lead in the musical life of his adopted country.

Claus Moser was another veteran whose acceptance speech of 1996 contained a moving tribute to Prince Albert and his German heritage. Lord Moser said: *"Inevitably, this occasion brings some personal thoughts to my mind. My origins, like those of Prince Albert, were in Germany. In fact, it is precisely 60 years since my family came to this country from Hitler's Germany. Ever since, I have felt deep gratitude to my parents for their decision to leave Germany, and to this country for enabling us to settle here for safe and happy lives. I have indeed been lucky to have lived most of my life here, even though the early Berlin years, including those in Hitler's time, no doubt left their mark. Memories of that Berlin childhood bring to mind some of the best of*

16 *Derek Hudson* and *Kenneth William Luckhurst*, The Royal Society of Arts 1754-1954, London 1954, 366, plate 36; *David Guy Charles Allan*, RSA: a chronological history, London 1999, 119, 124.

17 *Hudson* and *Luckhurst* (note 16), 352.

what Europe had to offer, as well as some of the most horrendous. The horrendous need no recalling tonight. The best perhaps does. In education, there were qualities which I imagine appealed to Prince Albert and to which I owe a great deal. No one doubted the overwhelming importance of education, the need for broad curricula focussing on the arts as much as on other subjects, and the respect due to teachers publicly and in the classroom.

Even more memorable was the cultural life. Germany's traditions in music, theatre, indeed all the arts, penetrated society. Academics, intellectuals and indeed ideas enjoyed high prestige. Culture was central to life and was expected to be progressive, daring and indeed provocative. No wonder Hitler chose the arts for early interference.

My love of the arts and respect for education no doubt stemmed from these early days and has stayed with me throughout my life.

It is natural that our thoughts this evening go back to the achievements of Prince Albert, in whose memory the Medal was established. His passions and influence covered the worlds of education, the arts, manufacture and trade, indeed all the areas which are so effectively pursued by the Society. The fruits of his efforts are all around us, for ever enriching our lives"[18].

The Duke of Edinburgh, who had himself benefited from the German educational vision, took an especial personal interest in the Society's work for industrial design and for the environment[19]. Both matters had a long pedigree in the Society's affairs, the former going back to the eighteenth century awards for textile design and to Prince Albert's determination to "*wed high art to mechanical skill*". The latter included pioneering attempts to prevent smoke nuisance in the early days of the industrial revolution, the improvement of housing for the labouring classes (another of Prince Albert's interests) and a major campaign to preserve English vernacular architecture in the 1920s. Albert medallists concerned with industrial design in this period have included Sir Nicolaus Pevsner (1975) and Sir Hugh Casson (1984). Among environmentalists we may count the awards to Sir Peter Scott (1970), the conserver of wildlife, to Sir John Betjeman (1973), poet and crusader against traffic pollution and suburban sprawl, to Baroness Warnock (1998) who had linked philosophy to education, ethics, human fertility and the environment, and to Tim Smit (2003), whose creation of the Eden project forms a part of the title of this paper.

Tim Smit's delight at receiving the Albert Medal reflects the prestige which the award still holds. As far as the people of the United Kingdom are concerned there is indeed only one Albert Medal, that of the Royal Society of Arts. In 1949 the Albert Medal for Gallantry was replaced by the George Cross and in 1971 all living recipients of the gallantry awards were invited to exchange their medals for the George Cross. So the Prince Consort's name lives on associated with an institution still served by his descendants. Surviving examples of the medal may be found in the Royal Archives at Windsor Castle, and in two National Trust houses: Craggside in

[18] Jnl RSA Dec 1996, 30.
[19] RSA, Prince Philip; Forty Years RSA President 1953-1992, London 1992.

Northumberland, home of Lord Armstrong, and Chartwell in Kent, Sir Winston Churchill's country residence. On account of their high quality gold and their historic association, such early examples are beyond the reach of most collectors. However, thanks to the generosity of a late twentieth century recipient, Mr Akio Morita (1982), the first medal of all – Sir Rowland Hill's – has been preserved and now remains in the safe keeping of the RSA. It may be hoped that other examples will be either collected or located, and that a record of all the awards, with the related documentary evidence, may be established. Such a record would not only be of value to historians but, to use Trueman Wood's heartfelt tribute, *"might fully be held to lend some further lustre to the reputation of the sagacious and benevolent Prince in whose memory it was founded"*[20].

APPENDIX

[R]SA Albert Medal Award Winners 1864-2003[21]

1864 Sir Rowland Hill, KCB, FRS (1795-1879) *"for his great services to Arts, Manufactures and Commerce, in the creation of the penny postage, and for his other reforms in the postal system of this country, the benefits of which have, however, not been confined to this country, but have extended over the civilised world"*.

1865 His Imperial Majesty Napoleon III (1808-1873) *"for distinguished merit in promoting, in many ways, by his personal exertions, the international progress of Arts, Manufactures and Commerce, the proofs of which are afforded by his judicious patronage of Art, his enlightened commercial policy, and especially by the abolition of passports in favour of British subjects"*.

1866 Michael Faraday DCL, FRS (1791-1867) *"for his discoveries in electricity, magnetism, and chemistry, which in their relation to the industries of the world have so largely promoted Arts, Manufactures, and Commerce"*.

1867 Sir W Fothergill Cooke (1806-1879) **and Sir Charles Wheatstone FRS** (1802-1875) *"in recognition of their joint labours in establishing the first electric telegraph"*.

20 *Trueman Wood* (note 1), 406.

21 Citations are those given in *Trueman Wood* to 1913 (note 1) and in the RSA Jnl thereafter. Styles and titles are those used by the medallists at the time of the reward. Biographical details may be found in the *Oxford Dictionary of National Biography*, *Chambers Biographical Dictionary*, *Who was Who* and *Who's Who*.
*Indicates a recipient of the Order of Merit, instituted by King Edward VII in 1902 on the lines of the order discussed by Lyon Playfair and Prince Albert in 1856.

1868 Sir Joseph Whitworth LLD, FRS (1803-1887) *"for the invention and manufacture of instruments of measurement and uniform standards by which the production of machinery has been brought to a state of perfection hitherto unapproached to the great advancement of Arts, Manufactures and Commerce".*

1869 Baron Justus von Liebig, Associate of the Institute of France, For.Memb.RS, Chevalier of the Legion of Honour, etc. (1803-1873) *"for his numerous valuable researches and writings, which have contributed most importantly to the development of food economy and agriculture, to the advancement of chemical science, and to the benefits derived from that science by Arts, Manufactures and Commerce".*

1870 Vicomte Ferdinand de Lesseps, Member of Institute of France, Hon GCSI (1805-1894) *"for services rendered to Arts, Manufactures and Commerce, by the realisation of the Suez Canal".*

1871 Sir Henry Cole KCB (1808-1882) *"for his important services in promoting Arts, Manufactures and Commerce, especially in aiding the establishment and development of International Exhibitions, the Department of Science and Art, and the South Kensington Museum".*

1872 Sir Henry Bessemer FRS (1813-1898) *"for the eminent services rendered by him to Arts, Manufactures and Commerce, in developing the manufacture of steel".*

1873 Michel Eugene Chevreul For.Memb.RS, Member of the Institute of France (1786-1889) *"for his chemical researches, especially in reference to saponifciation, dyeing, agriculture, and natural history, which for more than half a century have exercised a wide influence on the industrial arts of the world".*

1874 Sir William Siemens DCL, FRS (1823-1883) *"for his researches in connection with the laws of heat, and the practical applications of them to furnaces used in the Arts; and for his improvements in the manufacture of iron; and generally for the services rendered by him in connection with the economisation of fuel in its various applications to Manufactures and the Arts".*

1875 Michel Chevalier (1806-1879) *"the distinguished French statesman, who, by his writings and persistent exertions, extending over many years, has rendered essential services in promoting Arts, Manufactures and Commerce".*

1876 Sir George B Airy KCB, FRS, Astronomer Royal (1801-1892) *"for eminent services rendered to Commerce by his researches in nautical astronomy and in magnetism, and by his improvements in the application of the mariner's compass to the navigation of iron ships".*

1877 Jean Baptiste Dumas For.Memb.RS, Member of the Institute of France (1800-1884) *"the distinguished chemist, whose researchers have exercised a very material influence on the advancement of the Industrial Arts".*

1878 Lord Armstrong [William George Armstrong] **CB, DCL, FRS** (1810-1900) *"because of his distinction as an engineer and as a scientific man, and because by the development of the transmission of power – hydraulically – due to his constant efforts, extending over many years, the manufactures of this country have been greatly aided, and mechanical power beneficially substituted for most laborious and injurious manual labour".*

1879 Lord Kelvin [William Thomson] **OM, LLD, DCL, FRS** (1824-1907) *"on account of the signal service rendered to Arts, Manufactures and Commerce, by his electrical researches, especially with reference to the transmission of telegraphic messages over ocean cables".*

1880 James Prescott Joule LLD, DCL, FRS (1818-1889) *"for having established, after most laborious research, the true relation between heat, electricity and mechanical work, thus affording to the engineer a sure guide in the application of science to industrial pursuits".*

1881 August Wilhelm Hofmann MD, LLD, FRS, Professor of Chemistry in the University of Berlin (1818-1892) [formerly Georg Christian Wilhelm August Hofmann] *"for eminent services rendered to the Industrial Arts by his investigations in organic chemistry, and for his successful labour in promoting the cultivation of chemical education and research in England".*

1882 Louis Pasteur Member of the Institute of France, For.Memb.RS (1822-1911) *"for his researches in connection with fermentation, the preservation of wines, and the propagation of zymotic diseases in silkworms and domestic animals, whereby the arts of wine making, silk production and agriculture have been greatly benefited".*

1883 Sir Joseph Dalton Hooker KCSI CB MD DCL LLD FRS (1817-1911) *"for the eminent services which, as a botanist and scientific traveller, and as Director of the National Botanical Department, he has rendered to the Arts, Manufactures and Commerce by promoting an accurate knowledge of the floras and economic vegetable products of our several colonies and dependencies of the Empire".*

1884 Captain James Buchanan Eads (1820-1887) *"the distinguished American engineer, whose works have been of such great service in improving the water communications of North America, and have thereby rendered valuable aid to the commerce of the world".*

1885 Sir Henry Doulton (1820-1897) *"in recognition of the impulse given by him to the production of artistic pottery in this country".*

1886 Lord Masham (1815-1906) [Samuel Cunliffe Lister] *"for the services he has rendered to the textile industries, especially by the substitution of mechanical wool combing for hand combing, and by the introduction and development of a new industry – the utilisation of waste silk".*

1887 Her Majesty Queen Victoria (1819-1901) *"in commemoration of the progress of Arts, Manufactures and Commerce throughout the Empire during the fifty years of her reign".*

1888 Professor Hermann Louis Helmholtz For.Memb.RS (1821-1894) *"in recognition of the value of his researches in various branches of science and of their practical results upon music, painting and the useful arts".*

1889 John Percy LLD, FRS (1817-1889) *"for his achievements in promoting the Arts, Manufactures and Commerce, through the world-wide influence which his researches and writings have had upon the progress of the science and practice of metallurgy".*

1890 Sir William Henry Perkin FRS (1838-1907) *"for his discovery of the method of obtaining colouring matter from coal tar, a discovery which led to the establishment of a new and important industry, and to the utilisation of large quantities of a previously worthless material".*

1891 Sir Frederick Abel Bart, GCVO, KCB, DCL, DSc, FRS (1827-1902) *"in recognition of the manner in which he has promoted several important classes of the Arts and Manufactures by the application of Chemical Science, and especially by his researches in the manufacture of iron and of steel, and also in acknowledgement of the great services he has rendered to the State in the provision of improved war material, and as Chemist to the War Department".*

1892 Thomas Alva Edison (1847-1931) *"in recognition of the merits of his numerous and valuable inventions, especially his improvements in telegraphy, in telephony, and in electric lighting, and for his discovery of a means of reproducing vocal sounds by the phonograph".*

1893 Sir John Bennet Lawes Bart, FRS (1814-1900) **and Sir Henry Gilbert PhD, FRS** (1814-1900) *"for their joint services to scientific agriculture, and notably for the researches which, throughout a period of fifty years, have been carried on by them at the Experimental Farm, Rothamsted".*

***1894 Lord Lister FRS** (1827-1912) [Joseph Lister] *"for the discovery and establishment of the antiseptic method of treating wounds and injuries by which not only has the art of surgery been greatly promoted, and human life saved in all parts of the world, but extensive industries have been created for the supply of materials required for carrying the treatment into effect".*

1895 Sir Isaac Lowthian Bell Bart, FRS (1816-1904) *"in recognition of the services he has rendered to Arts, Manufactures and Commerce, by his metallurgical researches and the resulting development of the iron and steel industries".*

1896 Professor David Edward Hughes FRS (1831-1900) *"in recognition of the services he has rendered to Arts, Manufactures and Commerce, by his numerous inventions in electricity and magnetism, especially the printing telegraph and the microphone".*

1897 George James Symons FRS (1838-1900) *"for the services he has rendered to the United Kingdom by affording to engineers engaged in the water supply and the sewage of towns, a trustworthy basis for their work, by establishing and carrying on during nearly forty years systematic observations (now at over 3000 stations) of the rainfall of the British Isles, and by recording, tabulating and graphically indicating the results of these observations in the annual volumes published by himself"*.

1898 Professor Robert Wilhelm Bunsen MD, For.Memb.RS (1811-1899) *"in recognition of his numerous and most valuable applications of Chemistry and Physics to the Arts and to Manufactures"*.

1899 Sir William Crookes FRS (1832-1919) *"for his extensive and laborious researches in chemistry and in physics, researches which have in many instances developed into useful practical applications in the Arts and Manufactures"*.

1900 Henry Wilde FRS (1833-1919) *"for the discovery and practical demonstration of the indefinite increase of the magnetic and electric forces from quantities indefinitely small, a discovery now used in all dynamo machines; and for its application to the production of the electric search-light, and to the electro-deposition of metals from their solutions"*.

1901 His Majesty King Edward VII (1841-1910) *"in recognition of the aid rendered by His Majesty to Arts, Manufactures and Commerce, during thirty-eight years' Presidency of the Society of Arts, by undertaking the direction of important exhibitions in this country and the executive control of British representation at International Exhibitions abroad, and also by many other services to the cause of British Industry"*.

1902 Professor Alexander Graham Bell (1847-1922) *"for his invention of the Telephone"*.

1903 Sir Charles Augustus Hartley KCMG (1825-1915) *"in recognition of his services, extending over forty-four years, as Engineer to the International Commission of the Danube, which have resulted in the opening up of the navigation of that river to ships of all nations, and of his similar services, extending over twenty years, as British Commissioner on the International Technical Commission of the Suez Canal"*.

1904 Walter Crane (1845-1915) *"in recognition of the services he has rendered to Art and Industry by awakening popular interest in Decorative Art and Craftsmanship, and by promoting the recognition of English Art in the form most material to the commercial prosperity of the country"*.

***1905 Lord Rayleigh OM, DCL, ScD, FRS** (1842-1919) [3rd Baron, John William Strutt] *"in recognition of the influence which his researches, directed to the increase of scientific knowledge, have had upon industrial progress, by facilitating amongst other scientific applications, the provision of accurate electrical standards, the production of improved lenses and the development of apparatus for Sound Signalling at Sea"*.

1906 Sir Joseph Wilson Swan MA, DSc, FRS (1828-1914) *"for the important part he took in the invention of the incandescent electric lamp, and for his invention of the carbon process of photographic printing".*

***1907 The Earl of Cromer OM, GCB, GCMG, KCSI, CIE** (1847-1917) [Evelyn Baring] *"in recognition of his pre-eminent public services in Egypt, where he has 'imparted security to the relations of this country with the East, has established justice, restored order and prosperity, and, by the initiation of great works, has opened up new fields for enterprise".*

1908 Sir James Dewar MA, DSc, LLD, FRS (1842-1923) *"for his investigations into the liquefaction of gases and the properties of matter at low temperatures, investigations which have resulted in the production of the lowest temperatures yet reached, the use of vacuum vessels for thermal isolation, and the application of cooled charcoal to the separation of gaseous mixtures and to the production of high vacua".*

1909 Sir Andrew Noble KCB, DSc, DCL, FRS (1831-1915) *"in recognition of his long-continued and valuable researches into the nature and action of explosives, which have resulted in the great development and improvement of modern ordnance".*

1910 Madame Curie (1867-1934) *"for the discovery of Radium".*

***1911 The Hon Sir Charles Algernon Parsons KCB, LLD, DSc, FRS** (1854-1931) *"for his experimental researches into the laws governing the efficient action of steam in engines of the turbine type, and for his invention of the reaction type of steam turbine, and its practical application to the generation of electricity and other purposes".*

1912 Lord Strathcona and Mount Royal GCMG, GCVO, LLD, DCL, FRS (1820-1914) [Donald Alexander Smith) *"for his services in improving the railway communications, developing the resources, and promoting the commerce and industry of Canada and other parts of the British Empire".*

1913 His Majesty King George V (1865-1936) *"for nine years President, and now Patron of the Society, in respectful recognition of His Majesty's untiring efforts to make himself personally acquainted with the social and economic condition of the various parts of his Dominions, and to promote the progress of Arts, Manufactures and Commerce in the United Kingdom and throughout the British Empire".*

1914 Chevalier Guiglielmo Marconi LLD, DSc (1874-1937) *"for his services in the development and practical application of wireless telegraphy".*

***1915 Professor Sir Joseph John Thomson OM, DSc, LLD, FRS** (1856-1940) *"for his researches in physics and chemistry, and their application to the advancement of Arts, Manufactures and Commerce".*

1916 Professor Elias Metchnikoff For.Mem.RS (1845-1916) *"in recognition of the value of his investigations into the causes of immunity in infective diseases, which* *have led to important changes in medical practice, and to the establishment of principles certain to have a most beneficial influence on the improvement of public health".*

1917 Orville Wright (1871-1948) *"in recognition of the value of the contributions of Wilbur and Orville Wright to the solution of the problem of mechanical flight".*

1918 Sir Richard Tetley Glazebrook CB, ScD, FRS (1854-1935) *"for his services in the application of science to the industries of peace and war, by his work as Director of the National Physical Laboratory since 1899, and as Chairman of the Advisory Committee for Aeronautics".*

1919 Sir Oliver Joseph Lodge DSc, LLD, FRS (1851-1940) *"in recognition of his work as the pioneer of wireless telegraphy".*

1920 Professor A A Michelson (1852-1931) *"whose optical inventions have rendered possible the reproduction of accurate metric standards, and have provided the means of carrying out measurements with a minute precision hitherto unobtainable".*

1921 Professor John Ambrose Fleming FRS (1849-1945) *"in recognition of his many valuable contributions to electrical science and its applications, and specially of his original invention of the thermionic valve, now so largely employed in wireless telegraphy and for other purposes".*

1922 Sir Dugald Clerk KBE DSc, LLD, FRS (1854-1932) *"in recognition of his important contributions, both theoretical and practical, to the development of the internal combustion engine, who in its latter forms has rendered aerial navigation possible, and is also extensively employed in the motor car, and in the submarine and for many other purposes".*

1923 Major General Sir David Bruce KCB, DSc, LLD, FRCP, FRS (1855-1931) and **Colonel Sir Ronald Ross KCB, KCMG, DSc, LLD, MD, FRCS, FRS** (1857-1932) *"in recognition of the eminent services they have rendered to the Economic Development of the World by their achievements in Biological Research and the Study of Tropical Diseases".*

1924 HRH The Prince of Wales (1894-1972) *"in recognition of Services rendered to the Arts, Manufactures and Commerce as President of the British Empire Exhibition, and by his visits to the Dominions and Colonies".*

1925 Lieut-Colonel Sir David Prain CMG, CIE, ME, LLD, FRS (1857-1944) *"for the application of Botany to the development of raw materials of the Empire".*

1926 Professor Paul Sabatier Member of the Institute of France, For.Memb.RS (1859-1941) *"in recognition of his distinguished work in science and of the eminent services to industry rendered by his renowned researches in Physics and Chemistry, which laid the foundation of important industrial processes".*

1927 Sir Aston Webb GCVO, CB, PRA, PRIBA, FSA, LLD (1849-1930) *"for distinguished services to Architecture"*.

***1928 Sir Ernest (afterwards Lord) Rutherford OM, LLD, DSc, FRS** (1871-1937) *"for his pioneer researches into the structure of matter"*.

1929 Sir Alfred Ewing KCB, LLD, FRS (1855-1935) *"for his work in magnetism and his services to technical education"*.

1930 Professor H E Armstrong LLD, DSc, FRS (1848-1937) *"for his discoveries in Chemistry and his services to Education"*.

1931 HRH The Duke of Connaught and Strathearn KG (1850-1942) *"in grateful appreciation of his Presidency of the Society since 1911"*.

1932 Frank Brangwyn RA (1867-1956) *"for his services to decorative and commercial art"*.

1933 Sir William Lewellyn GCVO, PRA (1858-1941) *"for his encouragement of Art in Industry"*.

***1934 Sir Frederick Gowland Hopkins LLD, DSc, PRS** (1861-1947) *"for his researches in Biochemistry and the Constituents of Foods"*.

1935 Sir Robert Hadfield Bart, DSc, FRS (1858-1940) *"for his researches in Metallurgy and his services to the Steel Industry"*.

1936 The Earl of Derby KG (1865-1948) [Edward George Villliers Stanley] *"for the advancement of Commerce and Arts, especially in Lancashire"*.

1937 Lord Nuffield OBE (1877-1963) [William Richard Morris] *"for services to industry, transport and medical science"*.

1938 Her Majesty Queen Mary (1867-1953) *"in recognition of Her Majesty's unremitting interest in arts and manufactures, to the great benefit of British industry and commerce"*.

1939 Sir Thomas Holland KCSI, KCIE, DSc, LLD, FRS (1868-1947) *"for services to the mineral industries"*.

1940 John A Milne CBE *"for services to Industrial Art"*.

1941 President Franklin D Roosevelt (1882-1945) *"in recognition of his pre-eminent services to humanity as the fearless and resolute champion of the ideals of freedom and individual liberty"*.

***1942 Field Marshal J C Smuts CH** (1870-1950) *"Statesman. Soldier. Scientist. Philosopher"*.

1943 Sir John Russell OBE, DSc, FRS (1872-1965) *"for his researches and leadership in agricultural science and his services to husbandry in many lands"*.

1944 Sir Henry Tizard KCB, DCL, FRS (1885-1959) *"for his achievements in applying scientific principles to aeronautics and his services to advanced Technical Education"*.

***1945 Winston Spencer Churchill CH, FRS, MP** (1874-1965) *"whose foresight, faith and fortitude led free men to victory"*.

1946 Sir Alexander Fleming FRS (1881-1955) **and *Sir Howard Florey FRS** (1898-1968) *"for their services in the discovery and development of pencillin"*.

1947 Sir Robert Robinson MA, DSc, LLD, FRS (1886-1975) *"for his outstanding contributions to the advancement of organic chemistry"*.

1948 Sir William Reid Dick KCVO, RA (1879-1961) *"for National Memories in Living Stone"*.

***1949 Sir Giles Gilbert Scott** (1880-1960) *"Builder of a lasting heritage for Britain"*.

1950 Sir Edward Appleton GBE, KCB, MA, DSc, ScD, LLD, FRS (1892-1965) *"for outstanding services to science and industrial research"*.

1951 His Majesty King George VI (1895-1952) *"in respectful recognition of His Majesty's lifelong concern for the progress of industry and for industrial welfare"*.

1952 Air Commodore Sir Frank Whittle KBE, CB, FRS (1907-1996) *"for his development of the continuous-combustion gas turbine and jet propulsion"*.

***1953 Dr Edgar Douglas Adrian OM, MD, PRS** (1889-1977) *"for his outstanding contribution to neuro-physiology"*.

1954 Sir Ambrose Heal (1872-1959) *"for his services to industrial design"*.

***1955 Dr Ralph Vaughan Williams OM** (1872-1958) *"for his eminent service to music"*.

***1956 Sir Henry Dale OM, GBE, MD, FRS** (1875-1968) *"for eminent service to science, particularly physiology"*.

1957 Sir Christopher Hinton FRS (1901-1983) *"for his outstanding leadership in nuclear power development"*.

1958 HM Queen Elizabeth II (1926-) *"to mark Her Majesty's personal service to arts, manufactures and commerce, at home and abroad"*.

1959 Rt Hon Vincent Masey CH (1887-1967) *"for his distinguished encouragement of the arts and sciences"*.

1960 Sir Frederick Handley Page CBE (1885-1962) *"for his leadership in design and manufacture of aircraft"*.

1961 Professor Walter Gropius Dr.Ing, FIAA, Hon.RDI (1883-1969) *"for his contributions to architectural and industrial design".*

1962 Sir Gordon Russell CBE, MC, RDI, FSIA (1892-1980) *"for his services to industrial design".*

***1963 HRH The Duke of Edinburgh** (1921-) *"for distinguished merit in promoting arts, manufactures and commerce".*

***1964 Dame Ninette de Valois DBE** (1898-2001) *"for her services to the art of ballet".*

1965 Sir Leon Bagrit (1902-1979) *"for his work in the application of automation to industry".*

1966 Christopher Cockerell CBE (1910-1999) *"for his work in the invention and technical development of the Hovercraft".*

1967 Sir Edward Lewis (1900-1980) *"for his contribution to the electronics industry".*

1968 Sir Barnes Wallis (1887-1979) *"in recognition of his contributions to the development of aeronautical science and engineering".*

1969 Sir Allen Lane (1902-1970) *"for his contribution to publishing and education".*

1970 Sir Peter Scott (1909-1989) *"for his work in the conservation of wild life".*

1971 Sir William Glock CBE (1908-2000) *"for his outstanding services to music".*

***1972 Sir George Edwards OM CBE FRS** (1908-2003) *"for services to aeronautical science and aviation".*

1973 Sir John Betjeman CBE (1906-1984) *"for his contribution to poetry and the appreciation of architecture".*

1974 HM Queen Elizabeth the Queen Mother (1900-2003) *"in recognition for Her Majesty's outstanding support and patronage of the arts, manufactures & commerce".*

1975 Sir Nikolaus Pevsner (1902-1983) *"for his distinguished services to Britain's industrial heritage".*

***1976 Lord Olivier** (1907-1989) *"for his service to the Drama and the British Theatre".*

1977 Lord Robens PC (1910-1999) *"for his contribution to industrial progress in Britain".*

1978 Sir John Charnley (1911-1982) *"for his contributions to orthopaedic surgery".*

1979 Sir Robert Mayer (1879-1985) *"for his services to music, and in particular his generous and practical encouragement of young musicians and of young people learning to appreciate music".*

1980 Baroness Jackson of Lodworth (Barbara Ward) (1914-1981) *"for her work in the field of international co-operation in economic development"*.

***1981 Yehudi Menuhin Hon.KBE** (1916-1999) *"for his contribution to music"*.

1982 Akio Morita (b. 1921) *"for his contributions to technological and industrial innovation and management"*.

1983 Sir Arnold Hall (1915-2000) *"for his outstanding contributions to the aeronautical industry, and in particular to aeronautical engineering"*.

1984 Sir Hugh Casson KCVO RIBA RDI (1910-1999) *"for his contributions to art and design"*.

***1985 HRH The Prince of Wales** (1948-) *"for increasing recognition of the need for new – and often young – enterprise in industry, and for multiplying support, both corporate and private, for the arts"*.

1986 Sir Alastair Pilkington (1920-1995) *"for his outstanding contribution to industrial innovation"*.

1987 Dr Francis Crick FRS (1916-2004) *"for his contributions to molecular and cell biology"*.

1988 Sir Shridath Ramphal CMG QC (b. 1928) *"for his outstanding contributions towards accord within the Commonwealth, and his promotion of the worldwide concept of our inseparable humanity"*.

1989 Lord Sainsbury of Preston Candover *"for the outstanding contributions in the fields of business and the arts"*.

1990 Dr Jonathan Miller (b. 1934) *"for his outstanding contribution he has made to both the arts and science and the way he has brought both to a larger audience"*.

1991 Baroness Seear (1913-1997) *"for her distinguished contribution to public life in the spheres of industrial relations and the development of the principles of human resources management in industry"*.

1992 Lord Young of Darlington (b. 1915) *"for his impact and achievement in a multiplicity of fields, especially education, consumer affairs and social services"*.

1993 Paul Hamlyn (1926-2001) *"for his contribution to the arts, manufactures and commerce as a businessman, publisher and philanthropist"*.

1994 Sir Ernest Hall (1930-) *"for his charitable enterprise founded on his achievements as a financial analyst"*.

1995 Sir Adrian Cadbury (1929-) *"for his outstanding contribution to business and to corporate governance"*.

1996 Sir Claus Moser KCB CBE FBA (1922-) *"for his contribution to social sciences and commitment to education, music and the arts".*

1997 Sir Simon Rattle (1955-) *"for his outstanding contribution as a leading conductor and champion of orchestral involvement in a broad programme of education and community activity".*

1998 Baroness Warnock (1924-) *"in recognition of her national and international influence on the fields of education, ethics, human fertility, environmental issues and philosophy".*

1999 Professor Stephen Hawking (1942-) *"for making physics more accessible, understandable and exciting and opening the subject to a wider audience through his books and television programmes".*

2000 HRH The Princess Royal (1950-) *"for her many years of enthusiastically promoting and encouraging arts, manufactures and commerce in her visit programme and her dedicated work for charities".*

2001 Mary Robinson (1944-) *"for her outstanding work to advance corporate social responsibility in the global economy".*

2002 Tim Berners Lee (1955-) *"for his outstanding contribution to the World Wide Web".*

2003 Tim Smit CBE (1954-) *"for development of the Eden project which has broken new boundaries in tourism, ecology, education, enterprise and development partnership".*

Acknowledgements: Special thanks are due to Nicola Allen, Susan Bennett, Elizabeth Bonython, Franz Bosbach, Anthony Burton, John Davis, Christopher Eimer, Ben Francis, Frank James, Bernard Nurse, Alan Stevens and to staff members at the British Museum (Department of Coins and Medals), German Historical Institute, the Society of Antiquaries, the Institute of Historical Research, the Royal Society of Arts and the Richmond upon Thames Reference Library.

Autorenverzeichnis / List of Contributors

Dr. David G.C. Allan, 22a Headfield South, Twickenham TW2 7SS, U.K.

Prof. Dr. Franz Bosbach, Universität Bayreuth, Lehrstuhl für Geschichte der Frühen Neuzeit, D-95440 Bayreuth

Prof. Dr. Dr. Franz-Josef Brüggemeier, Universität Freiburg, Historisches Seminar, Lehrstuhl für Wirtschafts- und Sozialgeschichte, Werthmannplatz 1, D-79085 Freiburg

HD Dr. Jens Ivo Engels, Universität Freiburg, Historisches Seminar, Lehrstuhl für Wirtschafts- und Sozialgeschichte, Werthmannplatz 1, D-79085 Freiburg

Dr. Kai F. Hünemörder, Universität Göttingen, Graduiertenkolleg „Interdisziplinäre Umweltgeschichte“, Bürgerstraße 50, D-37073 Göttingen

Prof. Dr. Bill Luckin, Health and Social Studies, University of Bolton, Chadwick Campus, Bolton BL2 1JW, U.K.

Prof. Dr. Ian Simmons, University of Durham, Department of Geography, Science Laboratories, South Road, Durham DH1 3LE, U.K.

Prof. Dr. Christopher Smout, University of St Andrews, Institute for Environmental History, St Katharine's Lodge, The Scores, St Andrews, Fife, KY16 8AL, U.K.

Prof. Dr. Dieter Schott, Technische Universität Darmstadt, Institut für Geschichte, Residenzschloss, D-64283 Darmstadt

Dr. Fiona Watson, University of Stirling, Department of History, AHRC Research Centre for Environmental History, Stirling FK9 4LA, U.K.

Dr. Angus J.L. Winchester, Lancaster University, Department of History, Lancaster LA1 4YG, U.K.

Prinz-Albert-Studien / Prince Albert Studies

Herausgegeben von Franz Bosbach

11 **Deutschland und Rußland in der britischen Kontinentalpolitik seit 1815 / Germany and Russia in British Policy towards Europe since 1815**
Hrsg. von Adolf M. Birke und Hermann Wentker
1994. 217 Seiten. Gebunden.
€ 40,00 / sFr 69,00. ISBN 3-598-21411-1

12 **Politikverdrossenheit / Disillusioned with Politics**
Der Parteienstaat in der historischen und gegenwärtigen Diskussion. Ein deutsch-britischer Vergleich / Party Government in the Past and Present Discussion. An Anglo-German Comparison.
Hrsg. von Adolf M. Birke und Magnus Brechtken
1995. 136 Seiten. Gebunden.
€ 40,00 / sFr 69,00. ISBN 3-598-21412-X

13 **Kommunale Selbstverwaltung / Local Self-Government**
Geschichte und Gegenwart im deutsch-britischen Vergleich
Hrsg. von Adolf M. Birke und Magnus Brechtken
1996. 164 Seiten. Gebunden.
€ 40,00 / sFr 69,00. ISBN 3-598-21413-8

14 **Das Kreditwesen in der Neuzeit / Banking Systems in Modern History**
Ein deutsch-britischer Vergleich. Hrsg. von Franz Bosbach und Hans Pohl in Zusammenarbeit mit Gabriele Jachmich und Christoph Kampmann
1997. 183 Seiten. Gebunden.
€ 40,00 / sFr 69,00. ISBN 3-598-21414-6

15 **Künstlerische Beziehungen zwischen Eng-land und Deutschland in der viktorianischen Epoche / Art in Britain and Germany in the Age of Queen Victoria and Prince Albert**
Hrsg. von Franz Bosbach und Frank Büttner in Zusammenarbeit mit Michaela Braesel und Christoph Kampmann
1998. 230 Seiten. 70 Abbildungen. Gebunden.
€ 54,00 / sFr 93,00. ISBN 3-598-21415-4

16 **Imperium / Empire / Reich**
Ein Konzept politischer Herrschaft im deutsch-britischen Vergleich / An Anglo-German Comparison of a Concept of Rule
Hrsg. von Franz Bosbach und Hermann Hiery in Zusammenarbeit mit Christoph Kampmann
1999. 227 Seiten, 14 Abbildungen. Gebunden.
€ 54,00 / sFr 93,00. ISBN 3-598-21416-2

17 **An Anglo-German Dialogue**
The Munich Lectures on the History of International Relations. Hrsg. von Adolf M. Birke, Magnus Brechtken und Alaric Searle
2000. 286 Seiten, Gebunden.
€ 54,00 / sFr 93,00. ISBN 3-598-21417-0

18 **Prinz Albert und die Entwicklung der Bildung in England und Deutschland im 19. Jahrhundert / Prince Albert and the Development of Education in England and Germany in the 19th Century**
Hrsg. von Franz Bosbach, William Filmer-Sankey und Hermann Hiery unter Mitarbeit von Thomas Brockmann
2000. 256 Seiten. 25 Abbildungen. Gebunden.
€ 54,00 / sFr 93,00. ISBN 3-598-21418-9

19 **Religion und Politik in Deutschland und Großbritannien / Religion and Politics in Britain and Germany**
Hrsg. von Richard Bonney, Franz Bosbach und Thomas Brockmann
2001. 202 Seiten, 2 Abbildungen. Gebunden.
€ 54,00 / sFr 93,00. ISBN 3-598-21419-7

20 **Die Weltausstellung von 1851 und ihre Folgen / The Great Exhibition and its Legacy**
Hrsg. von Franz Bosbach und John R. Davis in Zusammenarbeit mit Susan Bennett, Thomas Brockmann und William Filmer-Sankey
2002 . 440 Seiten. 71 Abbildungen. Gebunden.
€ 78,00 / sFr 134,00. ISBN 3-598-21420-0

21 **Geburt oder Leistung? / Birth or Talent?**
Elitenbildung im deutsch-britischen Vergleich / The Formation of Elites in a British-German Comparison. Hrsg. von Franz Bosbach, Keith Robbins und Karina Urbach
2003 . 239 Seiten. 6 Abbildungen. Gebunden.
€ 58,00 / sFr 100,00. ISBN 3-598-21421-9

22 **Prinz Albert - Ein Wettiner in Großbritannien / Prince Albert - A Wettin in Britain**
Hrsg. von Franz Bosbach und John R. Davis
2004. 201 Seiten. 12 Abbildungen. Gebunden.
€ 58,00 / sFr 100,00. ISBN 3-598-21422-7

23 **Politische Memoiren in deutscher und britischer Perspektive / Political Memoirs in Anglo-German Context**
Hrsg. von Franz Bosbach und Magnus Brechtken
2005. 195 Seiten. Gebunden.
€ 58,00 / sFr 100,00. ISBN 3-598-21423-5

Bei Fragen zur Produktsicherheit wenden Sie sich bitte an:
If you have any questions regarding product safety,
please contact:

Walter de Gruyter GmbH
Genthiner Straße 13
10785 Berlin
productsafety@degruyterbrill.com